Springer Series in
CLUSTER PHYSICS

Springer
*Berlin
Heidelberg
New York
Hong Kong
London
Milan
Paris
Tokyo*

Physics and Astronomy | **ONLINE LIBRARY**

http://www.springer.de/phys/

Springer Series in
CLUSTER PHYSICS

Series Editors:
A. W. Castleman, Jr.   R. S. Berry   H. Haberland   J. Jortner   T. Kondow

The intent of the Springer Series in Cluster Physics is to provide systematic information on developments in this rapidly expanding field of physics. In comprehensive books prepared by leading scholars, the current state-of-the-art in theory and experiment in cluster physics is presented.

Mesoscopic Materials and Clusters
Their Physical and Chemical Properties
Editors: T. Arai, K. Mihama, K. Yamamoto and S. Sugano

Cluster Beam Synthesis of Nanostructured Materials
By P. Milani and S. Iannotta

Theory of Atomic and Molecular Clusters
With a Glimpse at Experiments
Editor: J. Jellinek

Metal Clusters at Surfaces
Structure, Quantum Properties, Physical Chemistry
Editor: K.-H. Meiwes-Broer

Clusters and Nanomaterials
Theory and Experiment
Editors: Y. Kawazoe, T. Kondow and K. Ohno

Quantum Phenomena in Clusters and Nanostructures
By S.N. Khanna and A.W. Castleman, Jr.

Water in Confining Geometries
By V. Buch and J.P. Devlin

Series homepage – http://www.springer.de/phys/books/cluster-physics/

S.N. Khanna   A.W. Castleman, Jr.

Quantum Phenomena in Clusters and Nanostructures

With 115 Figures and 8 Tables

 Springer

Prof. S.N. Khanna
Department of Physics, Virginia Commonwealth University
Richmond, VA 23284-2000, USA
snkhanna@saturn.vcu.edu

Prof. A.W. Castleman, Jr.
Eberly Distinguished Chair in Science
Evan Pugh Professor, Department of Chemistry and Physics
152 Davey Laboratory, The Pennsylvania State Univesity
University Park, PA 16802, USA
awc@psu.edu

ISSN 1437-0395
ISBN 3-540-00015-1 Springer-Verlag Berlin Heidelberg New York

Library of Congress Cataloging-in-Publication Data

Khanna, S. N.
Quantum phenomena in clusters and nanostructures / S.N. Khanna, A.W. Castleman, Jr. p. cm. - - (Springer series in cluster physics, ISSN 1437-0395)
ISBN 3540000151 (alk. paper)
1. Nanostructures. 2. Quantum theory. I. Castleman, Y. W. (Albert Welford), 1936- II. Title. III. Series.
QC176.8.N35 K43 2003
530.12- -dc21   2002034365

This work is subject to copyright. All rights are reserved, whether the whole or part of the material is concerned, specifically the rights of translation, reprinting, reuse of illustrations, recitation, broadcasting, reproduction on microfilm or in any other way, and storage in data banks. Duplication of this publication or parts thereof is permitted only under the provisions of the German Copyright Law of September 9, 1965, in its current version, and permission for use must always be obtained from Springer-Verlag. Violations are liable for prosecution under the German Copyright Law.

Springer-Verlag Berlin Heidelberg New York
a member of BertelsmannSpringer Science+Business Media GmbH

http://www.springer.de

© Springer-Verlag Berlin Heidelberg 2003
Printed in Germany

The use of general descriptive names, registered names, trademarks, etc. in this publication does not imply, even in the absence of a specific statement, that such names are exempt from the relevant protective laws and regulations and therefore free for general use.

Typesetting: LE-TeX Jelonek, Schmidt & Vöckler GbR, Leipzig
Cover concept: eStudio Calamar Steinen
Cover production: *design & production* GmbH, Heidelberg
Printed on acid-free paper   57/3141/YL - 5 4 3 2 1 0

Preface

One of the greatest triumphs of the last century was the development of quantum mechanics. The sharp lines and their position in the atomic spectra were the prime motivation that led to the Bohr model of the hydrogen atom, with the development of the Schrödinger equation ultimately providing the framework for understanding the properties of matter at the molecular level. In the experimental arena, technical developments over the past thirty years have enabled researchers to devise methods to fabricate structures that are so small that the energy levels of the systems present a discrete energy spectrum where the stability and the reactivity are determined by the nature of these electronic levels and the degree to which they are filled. In these miniature solids, clusters in some cases or assemblies that constitute nanoscale materials in others, the properties are controlled by the discrete quantum conditions associated with reduced size. Consequently, they are ideal systems for observing quantum effects. Indeed, numerous novel phenomena, e.g., magic numbers in the mass spectra, macroscopic quantum tunneling of magnetization, and quantum corrals, have provided novel examples for observing quantum effects at the nanoscopic scale. Of equal importance are the developments in experimental techniques that are now enabling researchers to unravel the quantum evolution by probing the excitation/relaxation dynamics of electronic states in real time. Theoretical techniques have developed to the extent of not only providing a fundamental understanding of the properties of nanoscale systems, but also having predictive capability. These unprecedented developments are going to guide scientific thinking and material designs well into the next century.

The field of clusters and nano-scale materials is itself a rapidly developing area of research for other reasons. This is due in part to developments in experimental techniques ranging from supersonic molecular beams, sol-gel formation, sputtering, ball milling, to ones used for the formation of micelles and polymers. Indeed, it has been possible to generate free and embedded clusters of controlled size and composition, nanoscale particles containing up to several million atoms, nanocomposites, and nanocrystalline materials. The enormous interest in these systems stems from the fact that they display a totally new class of physical, chemical, electronic, magnetic and catalytic properties attributable to the reduced size and related aspects of quantum confinement. Further, the properties change with size and composition. One can thus understand how the quantum character evolves from atoms to the bulk solids. These types of behavior offer not only challenges for their fundamental understanding, but also avenues for new technologies. For example, it is envisioned that small clusters/nano-structures could serve as the building blocks of a new class of cluster assemblies. Three examples are the fullerides made from fullerenes, cluster-assembled solids made from Met-Cars, and nanostructures made from quantum dots.

In this volume, we have collected a diverse set of topics to highlight the influence of quantum constraints at reduced sizes. It is difficult to cover all aspects in a single volume but we hope that the reader will find the limited material interesting and stimulating.

Richmond, VA; University Park, PA   *S.N. Khanna*
August 2002   *A.W. Castleman*

Contents

1 Cluster and Nanoscale Science: Overview and Perspective
A.W. Castleman, Jr., S.N. Khanna 1

1.1 Introduction 1
1.2 Cluster Types: Formation and Study 2
1.3 Theoretical Developments 3
1.4 Theme of the Book 3

2 Quantum and Classical Size Effects in Thermodynamic Properties
R.S. Berry 7

2.1 Introduction 7
2.2 Quantum Properties of Small Systems 10
2.3 Phases of Finite Systems 12
2.4 Phase Diagrams of Finite Systems 19
2.5 Conclusion 26
References 26

3 Photoelectron Spectroscopy
G. Ganteför 29

3.1 Introduction 29
3.2 Physics of Photoelectron Spectroscopy 31
3.3 Experimental Set Up 36
    3.3.1 Cluster Source 37
    3.3.2 Time-of-Flight Mass Spectrometer 37
    3.3.3 Laser 37
    3.3.4 Photoelectron Spectrometer 38
3.4 Results 39
    3.4.1 Example: Electronic Shells in Clusters of Simple Metals ... 39
    3.4.2 Example: The Size Dependence of the Band Gap 42
    3.4.3 Example: Chemical reactivity and electronic structure 45
    3.4.4 Example: Dynamics 48
3.5 Conclusion and Outlook 51

4 Quantum Tunneling of the Magnetization in Molecular Nanoclusters
R. Sessoli, D. Gatteschi, W. Wernsdorfer 55

4.1 Introduction 55
4.2 The Magnetic Anisotropy of Molecular Clusters 57
4.3 The Superparamagnetic Behavior.............................. 61
4.4 Longitudinal Field Dependence of the Relaxation Rate:
    The Stepped Hysteresis 63
4.5 Transverse Field Dependence of the Relaxation Rate:
    The Berry Phase 65
4.6 The Role of Dipolar Fields: The Non-Exponential Relaxation 69
4.7 The Role of the Nuclear Magnetic Moments: The Isotope Effect .. 74
4.8 Conclusions... 78
References 79

5 Magnetism of Free and Supported Metal Clusters
J.P. Bucher 83

5.1 Introduction 84
5.2 Simple Considerations 86
    5.2.1 Common Ideas on Magnetism 86
    5.2.2 Implications for Cluster Magnetism 87
5.3 The Stern-Gerlach Experiments 89
    5.3.1 Experimental Principles 89
    5.3.2 Magnetic Moment Measurements of Unsupported Clusters . 91
    5.3.3 A High Resolution Experiment: Nickel 92
    5.3.4 Temperature Dependence of the Giant Moments.......... 93
    5.3.5 Clusters of Non-Ferromagnetic Transition Metals 96
    5.3.6 Locked Moment Clusters and Spin Canting 97
5.4 Interpretation of the Beam Experiments....................... 99
    5.4.1 Magnetic Anisotropy................................... 100
    5.4.2 Coupling Between Magnetic Moment and Lattice:
          Deflection Profile 100
5.5 Growth Kinetics of Clusters on Surfaces 102
    5.5.1 Tuning the Clusters Density 102
    5.5.2 Island Shapes 104
5.6 Thermodynamic Growth Modes 106
    5.6.1 Growth Criteria....................................... 106
    5.6.2 Elastic and Structural Considerations 107
5.7 Organized Growth 108
    5.7.1 Incommensurate Modulated Layers 108
    5.7.2 Atomic Scale Template................................ 110
    5.7.3 Self Organization 112
    5.7.4 Periodic Patterning by Stress Relaxation 113

	5.7.5 Organization on Vicinal Surfaces	115
	5.7.6 Low Temperature Growth	115
5.8	Magnetic Properties of Nanostructures	116
	5.8.1 Isolated Clusters on Surfaces	117
	5.8.2 Interacting Islands and Chains	120
	5.8.3 The Two-Dimensional Limit	125
5.9	Conclusion and Outlook	130
References		132

6 Magnetism in Free Clusters and in $Mn_{12}O_{12}$-Acetate Nanomagnets
S.N. Khanna, C. Ashman, M.R. Pederson, J. Kortus 139

6.1	Introduction	139
6.2	Magnetic Moment of Free Clusters in Beams	140
6.3	Oscillatory Change in the Magnetic Moment of Ni_n Clusters upon H Adsorption	143
6.4	Quantum Tunneling and Atomic, Electronic and Magnetic Structure of $Mn_{12}O_{12}$-Acetate	146
6.5	Details of Theoretical Studies	149
6.6	Geometry and Electronic Structure of Isolated $Mn_{12}O_{12}$ Clusters	150
	6.6.1 Vibrational Frequencies of the Hexagonal Tower	151
6.7	Electronic Structure of $Mn_{12}O_{12}$-Acetate	152
6.8	Magnetic Anisotropy Energy	154
6.9	Conclusions and Extension to Fe_8 Nanomagnets	156
References		157

7 Size Effects in Catalysis by Supported Metal Clusters
A.A. Kolmakov, D.W. Goodman 159

7.1	Introduction	159
7.2	Methodology	160
	7.2.1 Thin Oxide Films as a Model Support	160
	7.2.2 Cluster Deposition: Density, Size and Control of Morphology	161
	7.2.3 Analytical Tools: Spectroscopy and Microscopy	164
7.3	Cluster Size and Reactivity	170
	7.3.1 Geometric Factors	170
	7.3.2 Electronic Factors	174
7.4	Examples of Size Effects in Cluster Reactivity	181
	7.4.1 Onset of the Reactivity of Au/TiO_2 with Metal-Nonmetal Transitions and the Dimensionality of Supported Clusters	181
	7.4.2 CO Dissociation on Structural Defects of $Rh/Al_2O_3/NiAl$ (110)	186
	7.4.3 CO Oxidation Over a Pt/MgO Monodispersed Catalyst	189

X   Contents

7.5   Concluding Remarks and Future Prospects 192
References 193

8   Delayed Ionization
E.E.B. Campbell, R.D. Levine 199

8.1   Introduction 199
8.2   Transition State Theory 200
8.3   Detailed Balance 203
8.4   Experimental: The Rate of Thermionic Emission 210
8.5   Experimental: Dynamics 213
8.6   Kinetic Model 217
8.7   Concluding Remarks 219

9   Cluster Dynamics: Influences of Solvation and Aggregation
Q. Zhong, A.W. Castleman, Jr. 223

9.1   Introduction 223
9.2   Charge-Transfer Reactions 224
      9.2.1   Photo-Induced Electron-Transfer Reactions 224
      9.2.2   Excited-State Proton-Transfer 226
      9.2.3   Excited-State Double Proton-Transfer 229
9.3   Caging Dynamics 231
      9.3.1   Caging Dynamics in Neutral Clusters 232
      9.3.2   Caging Dynamics in Anionic Clusters 234
9.4   Coulomb Explosion Process in Clusters 238
      9.4.1   Role of Clusters in the Coulomb Explosion Process 238
      9.4.2   Modeling of Coulomb Explosion Process 240
      9.4.3   Coulomb Explosion Imaging 243
9.5   Electronic Excitation, Relaxation and Ionization of Met-Cars 245
      9.5.1   Met-Cars: A Unique Molecular Cluster System 246
      9.5.2   Delayed Ionization 246
      9.5.3   Ultrafast Spectroscopy 248
9.6   Conclusion 251
References 252

10   Future Directions
A.W. Castleman, Jr., S.N. Khanna 259

Index 263

List of Contributors

C. Ashman
Departement of Physics
Virginia Commonwealth University
Richmond, VA 23284-2000
USA

R. Stephen Berry
Department of Chemistry
The University of Chicago
Chicago, IL 60637
USA

J.P. Bucher
Institut de Physique et Chimie
des Matériaux de Strasbourg
Université Louis Pasteur
23 rue du Loess
67037 Strasbourg
France

Eleanor E.B. Campbell
School of Physics
and Engineering Physics
Gothenburg University
& Chalmers University of Technology
41296 Gothenburg
Sweden

A.Welford Castleman, Jr.
Departments of Chemistry and
Physics
The Pennsylvania State University
University Park
PA 16802
USA

Gerd Ganteför
Department of Physics
University of Konstanz
78457 Konstanz
Germany

Dante Gatteschi
Department of Chemistry
University of Florence
50019 Florence
Italy

D.W. Goodman
Department of Chemistry
Texas A&M University
P. O. Box 30012
College Station
TX 77842-3012
USA

S.N. Khanna
Department of Physics
Virginia Commonwealth University
Richmond, VA 23284-2000
USA

A.A. Kolmakov
Department of Chemistry
and Biochemistry
University of California
Santa Barbara
CA 93103-9510
USA

J. Kortus
Center of Computational
Materials Science
Naval Research Laboratory
Washington DC 20375-5000
USA

R.D. Levine
The Fritz Haber Research Center
for Molecular Dynamics
The Hebrew University
Jerusalem 91904
Israel

M.R. Pederson
Center of Computational
Materials Science
Naval Research Laboratory
Washington DC 20375-5000
USA

Roberta Sessoli
Department of Chemistry
University of Florence,
50019 Florence
Italy

Wolfgang Wernsdorfer
Laboratoire L. Néel
CNRS
38042 Grenoble
France

Q. Zhong
Chemistry Division
Code 6111
Naval Research Laboratory
Washington, DC 20375
USA

1 Cluster and Nanoscale Science: Overview and Perspective

A.W. Castleman, Jr. and S.N. Khanna

1.1 Introduction

The field of cluster science has undergone an explosive growth in activity during the last decade, stimulated both by the large number of basic problems to which studies of clusters have provided new insights, as well as the vast array of applied areas to which clusters relate such as condensed matter physics and materials science. Elucidating from a molecular point of view the differences and similarities in the properties and reactivity of matter in the gaseous phase compared to the condensed state has been an overriding theme of many of the investigations, with both experimental and theoretical attention being directed to studies of structure, thermochemical properties, reactivity, and electronic excitation and relaxation dynamics. Particular interest in cluster research derives from the fact that aggregates of very small dimensions have properties that often differ significantly from those of the bulk material of which they are composed, and they typically display behavior that is not fully characterized as being a solid, liquid or gas. Indeed, clusters are often referred to as a new state of matter. As a result of their small physical size, a large fraction of the constituents of a cluster lies on the surface, and also constraints are imposed on their energy levels which frequently give rise to unique (quantum) effects. Furthermore, their study provides insights into the possibility of using clusters as building blocks for assembling new nanoscale materials in the future.

In order to bring these prospects to fruition, detailed investigations of their unique physical and chemical behavior are being actively pursued. Clusters are the ideal systems for observing quantum phenomenon since the properties are controlled by discrete quantum conditions associated with reduced size. The puzzle of the atomic spectra was the prime motivation for the development of the quantum mechanics. It was Niels Bohr who pointed out that the wave nature of the electrons forced them to occupy electron levels corresponding to the quantized angular momentum. The mass spectrum of the clusters has provided us with a similar situation. For example, the size distribution of Na clusters generated in beams are found to correspond to quantum states that can be obtained by applying Bohr's quantization rules to electrons moving inside the droplets in triangular and square orbits. In the area of magnetism, phenomenon like macroscopic quantum tunneling of magnetization, and quantum corrals have provided us with examples where quantum effects could be observed at a macroscopic scale. The developments in experimental techniques are enabling researchers to probe the evolution of the electronic wave function and atomic motions in real time. These developments not only provide us with new fundamental science but are also likely to take the field of material science into the next century.

1.2 Cluster Types: Formation and Study

Molecular clusters constitute an aggregated state of matter in which the individual constituents are held together by forces that are relatively weak compared to those responsible for the chemical bonding of the constituent molecules. Hence, by necessity, neutral clusters are usually formed by utilizing supersonic expansion techniques, which facilitate cooling and concomitant association reactions that lead to their growth. Supersonic expansion techniques including both continuous sources as well as pulsed jets are commonly used to produce beams of neutral clusters. In either the pulsed or continuous expansion methods, cooling of the beam is accomplished through the conversion of the random thermal energy of a high pressure source gas into a directed beam velocity.

In the case of metal and semiconductor clusters, the technique of laser vaporization has been especially valuable as a method for cluster production. However, high temperature condensation sources also have been found to be very valuable methods in the case of more volatile systems, especially in forming large alkali metal clusters for example.

Interrogating cluster properties has involved many different approaches, with mass spectrometry often playing a major role in identifying the cluster sizes which display a particular abundance, reactive behavior, or property. In many favorable cases, the combination of mass spectrometers with lasers has enabled the determination of optical spectra for clusters, leading to information on their structures. In more recent advanced methods of study, the use of fast lasers coupled with pump-probe techniques has enabled the dynamical aspects of clusters to be determined, including details of electronic excitation and relaxation processes. In yet other approaches, various flow reactors also have been used with mass spectrometer detection methods to learn about rates of reactions. The advantages of using mass spectrometers to determine size related properties are sometimes thwarted by concomitant fragmentation processes which often accompanies the ionization of neutral clusters when cations are the requisite species to be detected. On the other hand, often the fragmentation process can give rise to large abundances of clusters that are characteristically classified as "magic numbers". Study of the role and properties of these species is also a branch of this broad field of research, sometimes providing unique information about cluster structures or in some cases electronic characteristics.

1.3 Theoretical Developments

Despite successful determinations of the structures of certain clusters through spectroscopic methods, the number which have not been amenable to detailed study compared to those which have been successfully characterized, is enormous. Hence, not surprisingly, experimental cluster research depends heavily on theoretical support for determining the geometrical arrangement of atoms, their electronic structure and the magnetic properties. The first theoretical model that enabled cluster scientists to understand the magic numbers in Na_n clusters was the simple Jellium model where a cluster is defined as a sphere of uniform positive charge distribution and the electrons fill the electronic states for this potential. This simplified model not only explained the major peaks but with a little modification could account for the subpeaks in the mass spectrum and could provide general framework to discuss the polarizability, the optical spectra and other properties. It is amazing that despite the development of more sophisticated theoretical methods, this model continues to guide in the search of stable pure, mixed metal, and metal-semiconductor clusters.

While the Jellium model provides a convenient framework to discuss the global aspects, one is often interested in the detailed atomic and electronic structure. Over the past years, numerous theoretical methods have been used. They range from the semi-empirical methods such as CNDO (complete neglect of differential overlap), to the tight-binding methods, effective medium theories, quantum chemical techniques such as the configuration interaction (CI), and the density functional approaches. All these methods generally agree in showing that small clusters have compact structures that are different from the bulk arrangements. In addition to the geometrical structure, theoretical studies combined with experiments have provided information on many physical quantities that are inaccessible through direct experiments. For example, by combining the results of photoelectron experiments with those of theory, we can obtain considerable information about the magnetic moments of clusters as a function their size and composition. The theoretical developments have been critical to understand some the observed quantum effects such as resonant quantum tunneling of magnetization.

1.4 Theme of the Book

From a scientific perspective, it is fair to state that the development of quantum mechanics is ones of the greatest achievements of the past century. In the cluster field, among the most exciting findings of recent years are those which arise due to the confined sizes of particular clusters, which in turn derive from what is termed quantum confinement. In like manner, unique magnetic properties often emerge from the systems of restricted size and comprised of open shell atoms or molecules.

In the context of systems of finite size, one of the important considerations to arise in recent years pertains to effects on thermodynamic properties. In the second chapter of this book, Berry discusses the fundamental principles of classical thermodynamics which are used to treat bulk systems break, showing that they are invalid for ensembles of small systems. He explores ways in which the properties can be properly treated, leading to new insights into phase transitions, and systems in either phase equilibrium or undergoing phase changes.

Spectroscopic study of clusters provides invaluable information related to their geometric and often electronic structure. The third chapter, authored by Gantefor, demonstrates the depth of understanding that can be derived from studies employing photoelectron spectroscopy, where detailed information on mass selected systems can often be acquired. Recent developments are presented with attention to the new insights which have been derived for carbon-60, in comparison to other forms of carbon. Application to further elucidating the electronic properties of simple metals such as potassium, particularly in terms of the jellium model is also presented.

Three chapters deal with the magnetic properties of systems of small size, a topic of considerable current interest. The first of these deals with a consideration of quantum tunneling phenomenon in molecular nanoclusters. Here, Sessoli, Gatteschi, and Wernsdorfer consider the quantum tunneling of the magnetization as being a typical mesoscopic effect at the border between classical and quantum physics. Consideration is given to size domains where molecular clusters are small objects characterized by a small spin compared to a single domain particle, but significantly larger than the largest spin observable in atoms. Hence the magnetic properties of molecular clusters have a quantum character, but the macroscopic magnetization of the clusters are shown to exhibit bi-stability, hysteresis effects and super-paramagnetic behavior as the more classical single-domain particles.

A perspective of magnetism from an experimental point of view is provided by the chapter of Bucher, who presents an overview of Stern-Gerlach experiments on beams of metal clusters. The technique provides detailed information on the intrinsic magnetic moments of the clusters as well as on the relaxation processes involved when the free clusters cross a magnetic field. It is found that the observable projections of the magnetic moments onto the field axis scale with the magnetic field, cluster size and inverse vibrational temperature. The experimental findings are found to be in accord with the idea that the cluster moments are subject to rapid orientational fluctuations. A consideration of the magnetic moments of transitional metal and clusters of rare earths points to the fact that 3d and 3f ferromagnetism react quite differently to a confined geometry, due in part to a different relative importance of magnetic anisotropy and exchange energy. The chapter also treats the timely subject pertaining to critical magnetic phenomena involved during the growth and coalescence of surface deposits comprised of atoms and clusters.

Further insights into magnetic phenomena arise from the chapter by Khanna dealing with clusters and nanoparticles comprised of Fe, Co, and Ni, which in some cases have been found to be supermagnetic with large variations of moments with size as opposed to ferromagnetic behavior for the bulk material. In addition, clusters of non-magnetic elements like rhodium and antiferromagnetic solids like Mn are found to be ferromagnetic with large moments. The chapter presents a new approach to treat the systems based on a simplified exact method to calculate spin-orbit coupling in multicenter systems. New findings for systems composed of iron. Cobalt and mixed FeCo clusters containing up to 6 atoms are presented, together with findings for small Mn-oxide clusters which exhibit quantum tunneling of magnetic spins.

Another subject of considerable interest in the cluster field pertains to size effects on the catalytic behavior of supported metal clusters. In a chapter by Kolmakov and Goodman, the subject of "structure sensitivity" for metal nanoclusters dispersed on metal oxides is discussed in detail in terms of the fundamental understanding of the phenomena which has been derived from recent experimental results. Particular attention is devoted to reactions of CO on clusters of Au and Pd supported on transition metal oxide surfaces.

A phenomena of long standing interest in the cluster field if that of delayed ionization, first observed in metal oxide systems, and more recently in certain transition metal-carbon clusters, including Met-Cars, fullerenes, and perhaps even in a selected van der Waals cluster system. It is known that the phenomenon has some analogy to the bulk phase process of thermionic emission, but with some differences in part due to the finite sizes of the nanosystems under study in the cluster field. *Campbell and Levine* discuss current understanding of the processes in fullerenes in terms of the statistical processes arising due to the large density of states present. Particularly significant is the identification of the various time scales of ionization identified.

Determining the influence which solvation has on the dynamics of chemical reactions is another of the scientifically challenging problems in the field of chemical physics. In this area, there is particular interest in contrasting differences in the behavior and reactivity of ions as well as neutral species in the gaseous with those in the condensed phase, and studies of clusters at selectively increased degrees of aggregation offer the opportunity to explore the changes due to solvation. In this context, studies of selected chromophores excited within clusters, or cluster ions produced via selected ionization, and of the "ensuing solvation" that takes place within a cluster, shed light to the understanding of excitation processes in the bulk condensed state.

In a chapter on solvation and aggregation effects on reaction dynamics, *Zhong and Castleman* overview the role which femtosecond chemistry has played in elucidating the molecular details of various phenomena including electron and proton transfer as well as the production of multiply charged species, and Coulomb explosion in clusters. The subject of semiconductor excitation and relaxation dynamics is considered in the context of Met-Cars, including a brief discussion of their delayed ionization dynamics.

In terms of photoexcitation dynamics, processes of interest often occur on very short time scales and require special techniques for direct observation and consideration is given to reaction phenomena that can be unraveled via femtosecond pump-probe spectroscopy in the chapter on dynamics by Zhong and Castleman. New insights are acquired using the pump-probe technique pioneered by Zewail and coworkers, where herein attention is given to the influences of degree of aggregation on reactions of certain classes of systems, most particularly those involving hydrogen bonding. Particular consideration is given to a unique phenomena termed "Coulomb Explosion" which has been recently developed as a method of arresting and interrogating the course of fast reactions.

Other aspects of the field have developed along several lines including extensive efforts devoted to carbon clusters, to metal and semiconductor systems, rare gas and related van de Waals systems, as well as those composed of hydrogen bonded molecules. The latter are of primary value in elucidating the important process of proton transfer that pervades so many fields, and is a major focus of the work presented herein. The breadth of the field is demonstrated through discussing various proposed mechanisms by which high charge states may be developed, and also considering the nature of electron relaxation dynamics in metallic-like systems utilizing femtosecond pump-probe spectroscopy.

Another aspect of the field involves delayed ionization, a phenomena which is beginning to receive considerable experimental and theoretical attention. Various dynamical considerations are presented in the chapter by Zhong and Castleman, with an in depth treatment of current theoretical understanding of delayed ionization being given in Chap. 8 by Campbell and Levine.

The final chapter in this book deals with the future outlook for this promising area of research. We can expect many new and exciting insights to be gained in the area of cluster science focused in particular on the influence which quantum phenomenon has on the behavior of these systems of restricted size.

2 Quantum and Classical Size Effects in Thermodynamic Properties

R.S. Berry

The thermodynamics of small systems conform, of course, to the underlying, fundamental principles of thermodynamics. However many of the familiar relationships derived or rationalized for bulk matter from those principles are invalid for ensembles of small systems. This chapter first examines the conditions and constraints that enable us to use thermodynamics for small systems, and then explores ways in which the thermodynamic properties and behavior of small systems differ from their familiar counterparts for bulk matter. The greater part of the discussion focuses on phase changes and the properties of systems either in phase equilibrium or undergoing phase changes.

2.1 Introduction

In this introduction, we first outline how it is possible to apply thermodynamics to small systems. Then we go on to point out ways in which some of the dynamical properties of small systems differ from those of bulk systems. Recognition of these properties will be a key to understanding how and why the thermodynamic properties of small systems differ from their bulk counterparts, and how the properties of small systems transform with increasing size of system into bulk properties.

The first and very basic point is that thermodynamics, properly applied, is just as valid for describing small systems as it is for any other kind of system in or near equilibrium. However to use thermodynamics, one must follow Gibbs by framing every description in terms of the behavior of an ensemble of identical systems. "Identical" here means that all systems in the ensemble satisfy the same constraints that define that ensemble. For example, the ensemble might be microcanonical – in which all systems have the same energy – or canonical – with all systems at the same temperature – or grand canonical – with all systems at the same chemical potential. When one deals with clusters as the constituent systems, one must specify whether the clusters all have the same number of particles, may have all the possible sizes in a defined distribution, or may change size as they evolve. There are many useful ways to define ensembles to correspond to particular experimental conditions, real or hypothetical.

Often we study small systems by simulating them in molecular dynamics (MD) calculations. Provided the systems are ergodic, the time history of a single system in such a simulation would give a distribution of states, structures or whatever variability we are investigating that is equivalent to the distribution in the ensemble. The only conditions needed to maintain this equivalence in principle are that the simulation obeys precisely those constraints that define the ensemble and that the system is ergodic. In practice, the equivalence is easy to maintain because of a fundamental limitation of molecular dynamics simulations: the roundoff error in the computations accumulates to the extent that the simulated trajectories lose their mechanical reversibility in some tens of thousands of time steps, randomizing the simulated systems' exploration of their available phase space. In other words, we can safely use molecular dynamics simulations to infer thermodynamic properties.

Despite the validity of thermodynamics for describing small systems, the differences between small and bulk systems sometimes seem to make thermodynamics inapplicable to the small systems. Some of the best known concepts associated with thermodyamics fail us when we examine how small systems behave. One is the supposition that heat capacities must be positive. Another is the Gibbs phase rule, and with it, the concept of phase coexistence only along restricted sets of conditions such as the well-known pressure-temperature coexistence curves for two phases of homogeneous substances. We shall examine how these concepts fail to describe small systems, and how they evolve, as systems are made larger and larger, into the familiar forms of bulk thermodynamics.

The description of systems by ensembles defined by constraints is ubiquitous and useful. It is important to keep in mind, however, that we use the constraints to describe conditions that we suppose can be maintained for times long enough to permit an observer to study the system or an ensemble of such systems. Microcanonical ensembles are very useful for describing clusters isolated by rapid adiabatic expansions into environments of pressures so low that the clusters are out of contact with their environment for relatively long intervals. However the equilibrium we know best and experience in most conditions is thermal equilibrium, the condition of constant temperature. The only ensembles that we know describe distributions at thermal equilibrium are the Boltzmann distribution for classical systems, the Fermi-Dirac distribution for fermions, particles of half-integral spin, and the Bose-Einstein distribution for bosons, for particles of integral spin. At sufficiently high temperatures, both the Fermi-Dirac and Bose-Einstein distributions become indistinguishable from the Boltzmann distribution. Under many conditions, particularly of constant pressure, temperature and chemical potential, the grand canonical ensemble becomes the appropriate distribution to describe the equilibrium state. This is specifically the case if the composition of the ensemble includes systems with different compositions, e.g. clusters of different sizes in dynamic equilibrium.

Another aspect of ensembles that needs recognition is the disparity between the idealized constraints that define theoretical and computational ensembles and the constraints of the real, experimental ensembles studied in the laboratory. One typical example is the one just cited, the use of the microcanonical ensemble to describe particles effectively isolated for some observable time by a fast expansion. The ideal ensemble is monoenergetic; the real ensemble does consist of particles effectively isolated and unable to exchange energy for the same interval, but the particles in the real system of course have some distribution of energies. Likewise, and we shall examine an interesting example of this later in this chapter, a theoretical description of an ensemble of clusters that absorb energy from a burst of radiation may assume that the energy is thermalized within the clusters before any further process occurs, but the time scale for evaporation may, in principle, be no longer than (or even briefer than) the time scale for equilibration and thermalization of the vibrational energy.

One useful way to construct the quasi-equilibrium that sometimes characterizes systems especially small systems that have not yet reached true equilibrium, depends on the condition of different sets of degrees of freedom coming relatively quickly to a local equilibrium, yet between those sets, equilibrium may be achieved only on a much longer time scale. For example, in gases at pressures of order 1 torr or more, translations and rotations come to mutual equilibrium far faster than these degrees of freedom equilibrate with vibrations. Degrees of freedom associated with establishment of chemical equilibrium via intermolecular reactions typically takes longer still. Sometimes some subset of vibrational degrees of freedom may come to a quasi-equilibrium rapidly but equipartition their energy with other vibrational degrees of freedom rather slowly. In short, sometimes systems establish *local* equilibria for observable intervals, well before they come to true equilibrium.

The foregoing discussion has introduced the notion of time scales in the context of choosing an ensemble that describes the conditions of a system on the time scale of the observations used to study that system. Time scales and time scale considerations lurk throughout thermodynamics. For example, the universally invoked "reversible adiabatic" process is, in a naive interpretation, an oxymoron. If the process is so slow that it is reversible in every respect, then the system cannot be so well insulated that heat transfer does not occur. However the concept is extremely useful and entirely valid in a vast realm of situations. All we need require is that the time scale for relaxation to equilibrium within the system is short relative to the time scale for heat exchange with the surroundings [1]. This kind of time scale separation is so natural and so widely – but tacitly – accepted that we use it constantly as we apply thermodynamics in macroscopic contexts. One of the fascinations of the thermodynamics of small systems is the way such time scale separations are frequently invalid. The time scale for a small system to exchange a significant fraction of its energy with its surroundings may be very similar to the typical time for the degrees of freedom within the system to come to equilibrium.

The differences among ensembles and the issue of whether time scales are separable or not leads to the altogether plausible (and valuable) concept of nonequilibrium ensembles to describe realistic experimental situations. It is quite possible to have an ensemble in the laboratory that is well defined by conditions of an experiment and persistent long enough to permit observations, and to have the properties of the ensemble well enough defined to justify using an ensemble description. A particularly vivid example is the "evaporative ensemble" described by Klots [2]. In such an ensemble, the particles have presumably emerged as isolated units, but with an initial distribution of energies that permits many of the particles to lose component atoms or molecules by evaporation, on a time scale short relative to the transit time to the region where the particles are observed. With each evaporation, a composite particle or cluster loses energy. Consequently the next evaporation typically occurs more slowly than the first. Eventually, the evaporations are slower than the transit and observation time, so that the distribution under those conditions is effectively "frozen in." Such a distribution can be constructed fairly precisely, if the thermodynamic properties of the individual component particles and the binding energies of those component atoms or molecules are known. The evaporative ensemble is certainly not that of a system in equilibrium, but it does describe ensembles very much of a kind observed in experiments.

Still another characteristic of small systems is the relative size of fluctuations. As with all ensembles that have constant properties with distributions of values, the fluctuations and distributions in those quantities describing the individual systems that are not fixed vary as $N^{-1/2}$ where N is the number of particles comprising each of the individual systems. Hence the energies of the bulk systems in an ensemble at constant temperature have a very narrow spread around their mean, relative to the total energy of those systems. However the spread of energies around the mean for an ensemble of 20-atom clusters at constant temperature may be comparable to the total mean energy of each cluster. These large fluctuations imply that the differences between different ensembles can be extremely important for small systems, even though the differences between ensembles for bulk systems are frequently so small that we may use whichever distribution is most convenient for us.

2.2 Quantum Properties of Small Systems

Quantum properties of small systems manifest themselves in two ways. First, and most obvious, quantum effects appear in systems whose dimensions are small enough to make evident the wavelike properties of the matter comprising the system. Second, quantum effects appear in small systems at low temperatures, at which only a few quantum states (or perhaps only a single quantum state) may be populated significantly. There is experimental evidence that the bosonic atoms of alkali metal clusters at very low temperatures drop into their lowest common state, forming a Bose-Einstein condensate [3]. More recently, helium clusters have been shown to display a superfluid state, much like that of bulk helium [4,5]. The key evidence of superfluidity appears in the rotational spectra of foreign molecules trapped inside the helium clusters. Their rotations are essentially identical with those of the free molecules, implying that their rotational motion is completely unimpeded by any sort of collision with the helium atoms surrounding them.

The first effect can be thought of in terms of what happens when a particle of low mass is confined to a small volume; its energy levels become like those of a particle in a box, with spacings inversely proportional to the (square of) each of the linear dimensions of the box. For electrons and electronic contributions to the thermodynamic properties of a nonmetallic cluster, this effect is indistinguishable from its counterpart in ordinary molecules. The system exhibits the same kind of discrete electronic states that are so well known for molecules, with typical level spacings of order 2–4 eV, corresponding to excitations by visible or ultraviolet light. The dimensions to which a valence electron is confined in an insulating or semiconducting cluster are typically of order of the interatomic distance, like the electrons in a typical chemical bond.

Metal clusters behave rather differently, especially clusters of perhaps 50 or more atoms. In these systems, the valence electrons are confined only to a "box" with the dimensions of the cluster itself. This means that the energy levels available to the valence electrons are very closely spaced. They are so closely spaced that the clusters are often thought of as being metallic, with the valence electrons acting as conduction electrons.

If the valence electrons of a metal cluster can move relatively freely throughout the cluster's volume, then it is natural to ask whether the set of valence electrons might not show a shell structure, somewhat analogous to the shell structure of an atom or a nucleus. In fact, electronic shell structure was predicted [6–9] and manifested experimentally via abundance distributions [10–12], notably for alkali clusters. The shell model is somewhat less reliable as a predictor of properties of metal clusters with more than one valence electron per atom, such as aluminum clusters [13].

The shell model describes a metallic phase; clusters of only a very few atoms of alkali atoms are indeed rather well described by this model. However we would not expect a cluster of three or four atoms of magnesium, with closed $3s^2$ atomic subshells, to behave like tiny bits of metal, and in fact small clusters of alkaline earth atoms are bound by van der Waals forces, not by metallic binding forces. Small mercury clusters, too, composed of atoms with two outer s-electrons, are held together by van der Waals forces. We can therefore expect these systems to exhibit a transition from an insulating phase to a conducting, metallic phase as the size of the cluster increases. This subject has been reviewed recently by Johnston [14]. We might also expect a high-temperature metallic phase to appear in clusters just below the size at which the ground state becomes metal-like. The size-dependent transition has indeed been studied, initially theoretically for beryllium clusters [15], and experimentally with mercury clusters [16].

This is a venerable subject in the context of bulk solids [17], but when one considers the metal-insulator transition in clusters, both size and structure play roles, making the subject even richer than for bulk materials. The alkaline earths were studied in more detail theoretically [18,19], and the coinage metals, experimentally via their photoelectron spectra [20]. This technique was used more recently to study the metal-insulator transition in clusters of aluminum atoms [21]. The subject is complicated because some phenomena that appear together need not be equivalent for small systems. For example the density of states may be relatively sparse, relative to the mean thermal energy $k_\mathrm{B}T$ yet the one-electron states may be highly delocalized throughout the cluster. In a bulk metal, these two properties occur together. Hence there is always a question for clusters of what criterion would be the most apppropriate to distinguish the metal-insulator transition. Probably the most appropriate is the high-frequency electronic conductivity, taken at a frequency just high enough that the mean free path of the field-driven electron is a bit shorter than the linear dimension of the cluster.

2.3 Phases of Finite Systems

Small systems, in one sense, do not have phases; they do not exhibit the kinds of macroscopically static equilibrium that we see in a glass of water at thermal equilibrium and in equilibrium with the ambient pressure of water vapor in the room in which it stands. However in another sense, probably a broader and more useful sense, they may and often do exhibit forms that we ought to call phases. The ordinary solid and liquid forms are relatively easy to see. There is considerable evidence, experimental and theoretical, that at low temperatures, almost all kinds of clusters behave just as we expect solids to behave. For example, electron diffraction measurements of cold clusters show patterns characteristic of rigid, well-defined structures [22–38]. The one definite exception to this is clusters of helium atoms which, as mentioned previously, transform from liquid to superfluid at very low temperatures. No solid form is known for clusters of helium. Whether hydrogen clusters could exhibit a superfluid phase is still uncertain.

Clusters of many kinds of atoms tend to take on solid structures different from the lattice structures that characterize bulk crystals; the most common of these is icosahedral. The atoms of such clusters arrange themselves to produce regular icosahedra if there are precisely enough of them to make such structures. These are the "magic numbers" for icosahedra: 13, 55, 137, ... [39]. Each magic number corresponds to another completed outer shell; the 55-atom cluster has a central atom with two complete icosahedral layers around it, for example. Clusters of rare gas atoms (apart from helium) tend very strongly to take on icosahedral structures; clusters of most sizes below several hundred atoms, at least, have lowest-energy structures based on icosahedral shells. The few exceptions to this generalization, for example Ar_{38} and Ar_{75}, have far more icosahedrally-based structures than close-packed, the forms of their global-minimum structures [40,41]. Hence we can generalize that icosahedral structures *dominate* the stable forms of these clusters up to at least hundreds of particles.

Clusters of some other substances also tend to have polyhedral structures based on icosahedra. For example C_{13} has an icosahedron as its global minimum [42]. On the other hand, many other metal clusters have structures very different from icosahedra [15,43–45]. These, more typical of metal clusters, are very specific to the size of the clusters and to the nature of the constituent atoms. Neutral, positively-charged and negatively-charged clusters of the same number of atoms may have very different structures in their most stable states.

Most simulations of phase changes of clusters have been carried out at zero pressure, i.e. with no vapor present. These yield the coexistence range of temperature for that special pressure. A very few studies have been done with pressure as a variable. One was the relatively early simulation of sodium clusters by Holian et al. [46]. Another was that by Cheng et al. for a general approach to phase diagrams for clusters [71], which we discuss in the next section, and for the application of this approach to rare-gas (Lennard-Jones) clusters.

The greatest single difference between the thermodynamics of phase equilibrium for bulk matter and for small particles is a qualitative difference that is a simple consequence of the difference between small and large numbers. In a large system, we observe two phases in (macroscopically) static equilibrium only when the mean chemical potentials of the two phases are equal: $\Delta\mu = 0$ for transfer of any material from one phase to the other. By contrast, two *components* may coexist in any relative proportion in a bulk system, so long as both components can exhibit local stability. We commonly write an equlilbrium constant for any two components, A and B, as $K = [A]/[B] = \exp(-\Delta F/k_B T)$, where ΔF is the free energy difference between the two components. We can write the same kind of equilibrium constant for two phases in equilibrium, e.g. liquid and solid: $K_{\text{phase}} = [\text{liquid}]/[\text{solid}] = \exp(-\Delta F_{\text{liq–sol}}/k_B T) = \exp(-N\Delta\mu/k_B T)$, where N is the number of atoms or molecules comprising the entire system and $\Delta\mu$ is the mean chemical potential difference, per particle, between the two phases. Thus, even if $|\Delta\mu/k_B T|$ is very, very small, e.g. only 10^{-10}, but $N = 10^{20}$, then $K_{\text{phase}} = \exp(\pm 10^{10})$. In other words, the equilibrium constant is either enormous, effectively infinite, or infinitesimal, effectively zero, unless $\Delta\mu$ is precisely zero. That all N of the particles contribute to the interactions that stabilize a phase makes the leverage of the macroscopic value of N so great that the unfavored phase is always so unfavored that it would be present in unobservably small amounts in any equilibrium system. However we know very well that *local* stability of an unfavored phase may persist to temperatures and pressures far from the point of equal chemical potentials. We are familiar with superheated and supercooled (or undercooled) water, for example. It is the leverage of the large value of N, not the inherent lack of local stability, that makes unfavored phases unobservable in equilibrium.

Now let us turn to the phase behavior of small systems, consisting of perhaps 10 up to 10^6 or 10^8 particles. For these systems, in the range of conditions in which $\Delta\mu/k_BT$ is of order 10^{-6}, $N\Delta\mu/k_BT$ is small, up to only about unity. Hence the equilibrium constant for phase equilibrium, K_{phase} is itself roughly between e and $1/e$, so we can be confident that for ensembles of small clusters in equilibrium, we can expect to observe two or more phases in thermodynamic equilibrium under *ranges* of conditions, such as within ranges of temperature and pressure. The only strict condition for observability of a phase is that it must correspond to a local minimum in the free energy, with respect to some suitable order parameter, under the conditions of other thermodynamic variables such as temperature and pressure being constant.

Thus the phase rule is appropriate for bulk matter, for which we can expect to see observable amounts of two phases in equilibrium only when $\Delta\mu/k_BT = 0$. This means that only when that condition is met can the equations of state for two phases be satisfied. To satisfy the equations of state for two phases simultaneously while their chemical potentials are equal means that the two thermodynamic variables of the system cannot be independent; only one, e.g. temperature, may vary independently, while the other, e.g. pressure or volume, must change to maintain the equality of chemical potentials while the equations of state are satisfied.

Clusters are different; there is no constraint that the chemical potentials of the coexisting phases be equal. The conditions for coexistence of phases are just the same as those for coexistence of components. The number N of particles comprising a system, i.e. a cluster, is of the same magnitude as the number comprising an element of a component, i.e. the number of atoms in a typical molecule. The different phases of a cluster are precisely analogous to the isomers of a molecular component. In fact, with clusters, *we lose the distinction between phase and component*!

An important aspect of the difference in phase equilibria of clusters and of bulk matter is again a consequence of the small number of particles in a cluster and the indistinguishability of a phase and a component for a cluster. The equilibrium between coexisting phases of bulk matter is, at least at a macroscopic scale, a static equilibrium. There is no macroscopic scale for clusters; consequently whenever they may exist in more than one phase in an

2 Quantum and Classical Size Effects in Thermodynamic Properties   15

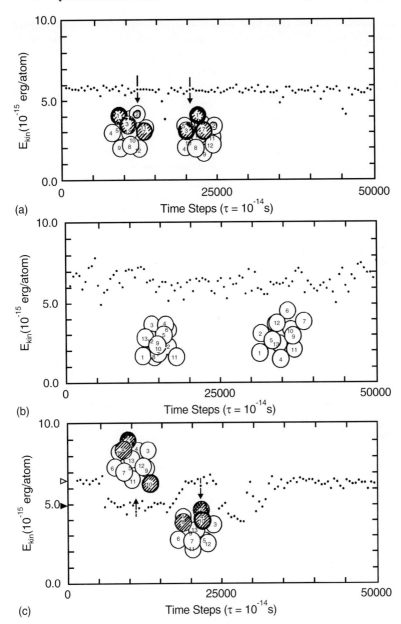

Fig. 2.1. Three time histories of the mean temperature, i.e. mean kinetic energy per particle, for Ar_{13} at three energies; (**a**) a low energy, at which the system is only solid; (**b**) a high energy, at which the system is only liquid; (**c**) an intermediate energy, at which the system can exhibit both solid and liquid forms, an energy at which an ensemble of such clusters would show the two phases in equilibrium, albeit a dynamic equilibrium

equilibrium ensemble, they pass back and forth, randomly, between or among those phases. Typically these passages occur on a scale of many picoseconds or nanoseconds. Figure 2.1 shows three time histories of the effective temperature, i.e. the mean kinetic energy per particle, for a cluster of Ar_{13}. All three were molecular dynamics simulations at constant energy. The first was done at an energy at which the cluster can only be solid. The second, Fig. 2.1b, was done at an energy at which the cluster is liquid, as the snapshots show. The third, Fig. 2.1c, was run at an energy in the range of coexistence of solid and liquid clusters. Clearly the system shows a bimodal distribution of mean kinetic energies or effective temperatures. Sometimes it is a hot solid, with very low potential energy and consequent high kinetic energy, and sometimes it is a cold liquid, with very little kinetic energy, since the liquid-like parts of the potential surface are high in energy. The fractions of time spent in each phase correspond to the fractions that define the equilibrium constant.

Figure 2.2 illustrates the shift in population with temperature, again for an Ar_{13} cluster but this time, from constant-temperature molecular dynamics simulations. The plots show the distributions of short-time mean energies (averages over ca. 500 time steps), displaying the passage from a unimodal distribution for solid only through a succession of bimodal distributions, finally to a unimodal distribution for the liquid only.

Figure 2.3a shows a similar time history for the cluster $(KCl)_{32}$. Here the vertical axis is the "quenched" potential energy, the potential energy of the minimum around which the system is vibrating at each instant at which a record is made [69]. The system passes between the low-energy rocksalt structure and high-energy, liquid-like structures. Figure 2.3b shows the proportions of solid and liquid for this cluster as a function of temperature.

A consequence of our logic so far is this: so long as they maintain local stability, three or more phases of clusters may exist in equilibrium over bands of temperature and pressure. There is clear evidence of this from simulations. Clusters of at least 45 argon atoms can exhibit a phase-like form, often called the "surface-melted" phase, in which the traditional diagnostic indicators–diffusion coefficients, fluctuations of interparticle distances, extent of ordered geometry–show that the outer layer behaves like a liquid but the inner core remains solid. This was first recognized by Nauchitel and Pertsin [70]; later, it was found that the liquid-like character inferred from numerical indicators and snapshots was somewhat misleading [71]. In fact, the "surface-melted" state has almost all the atoms of the outer layer undergoing very large-amplitude, very anharmonic motions, but in a very organized, collective way, around a well-defined polyhedral structure. However a few of those outer atoms, about one in thirty, moves out of the outer layer and moves, in a slightly restricted way, around the surface, as a "floater". Every few thousand vibrational periods, a floater exchanges with one of the surface atoms, so all the surface atoms get to be floaters and the outer layer of atoms does slowly achieve the permutational equivalence we identify with a liquid.

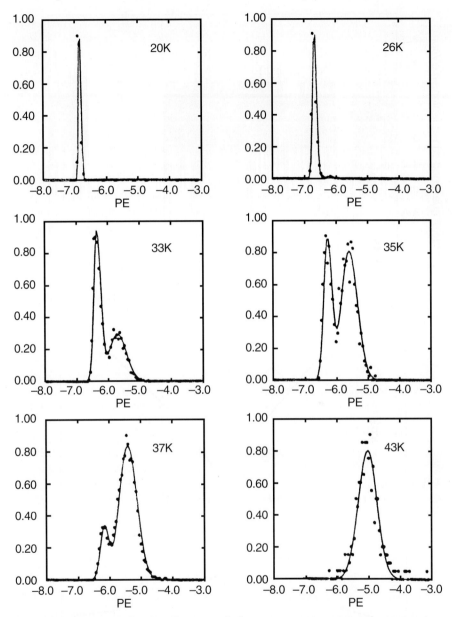

Fig. 2.2. Graphs of the distributions of short-term mean energy (averages over ca. 500 time steps) of an Ar_{13} cluster taken from constant-temperature molecular dynamics simulations. The first and last figures show the unimodal distributions of solid and liquid, respectively; the others show a progression in bimodal distributions from predominantly solid to predominantly liquid

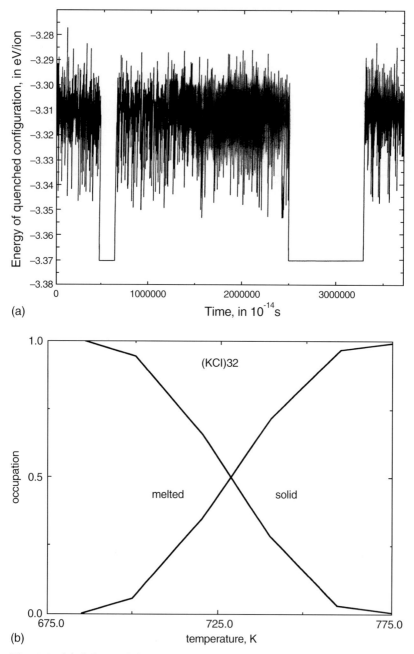

Fig. 2.3. (a) A "quench history" or time history of the energies of the local minima around which a $(KCl)_{32}$ cluster is vibrating, showing the bimodal distribution between the low-energy solid and the high-energy liquid; (b) the fractions of liquid and solid for this cluster, as functions of temperature

This conforms to one of the models for what is generally called, imprecisely, "surface melting" of bulk materials [72]. Simulations show that the surface-melted form can coexist over a range of temperatures with the liquid and solid forms [73,74]. Figure 2.4a shows this phenomenon in a "quench history" of Ar_{55}. The distribution of quenched potential energies is clearly trimodal for the temperature at which this simulation was done. Figure 2.4b shows the proportion of solid, surface-melted and liquid systems in an equilibrium ensemble in the temperature range of the coexistence of all three phases. This coexistence was explored further by Calvo and Labastie [75]. A study of clusters of nickel atoms could also be interpreted in terms of the existence of a surface-melted phase [76].

There are many other forms that clusters may assume that have all the character of phases. Because of the observability of thermodynamically unfavored phases, it is possible and even likely that clusters will exhibit phases that can never be seen in bulk matter in equilibrium. Such minority phases have been predicted and probably even seen for molecular clusters, e.g. the partially orientationally-ordered monoclinic phase of clusters of TeF_6 [77–79].

The general subject of phase behavior of clusters was reviewed recently [80].

2.4 Phase Diagrams of Finite Systems

Phase diagrams of bulk systems are well-known subjects of traditional courses in thermodynamics and frequent objects of study, especially experimentally. Their utility and simplicity have made them part of the engineer's and materials scientist's bag of tools. Underlying these diagrams is the Gibbs phase rule, $f = c + 2 - p$, where f is the number of degrees of freedom, c is the number of chemical components, and p is the number of phases present, i.e. the number of equations of state that must be satisfied simultaneously. The underlying assumptions, that phases and components are clearly distinguishable, and that only a negligible fraction of the matter in the system is in interfaces between phases, are well-grounded for bulk matter, but, as we have seen, are not necessarily valid for clusters and other highly-divided forms of matter. The phase rule tells us that two phases may coexist in equilibrium only along a curve, corresponding to a single degree of freedom, representable for example as a function $p_{\text{eq}} = f_{\text{eq}}(V)$. The condition for this curve is the equality of the mean chemical potentials of the two coexisting phases, $\Delta\mu = 0$.

Given that, as we have seen, clusters exhibit ranges of coexisting phases, rather than sharp curves of coexistence where $\Delta\mu = 0$, we need a way to portray the phase coexistence of clusters that is richer and more flexible than the conventional planar graph of coexistence loci. Specifically, we would like a way to portray the equilibrium ratio of the amounts of two (or more) phases in equilibrium, as a function of temperature and, if possible, of pressure. In this section, we describe two kinds of phase diagrams, one that exhibits

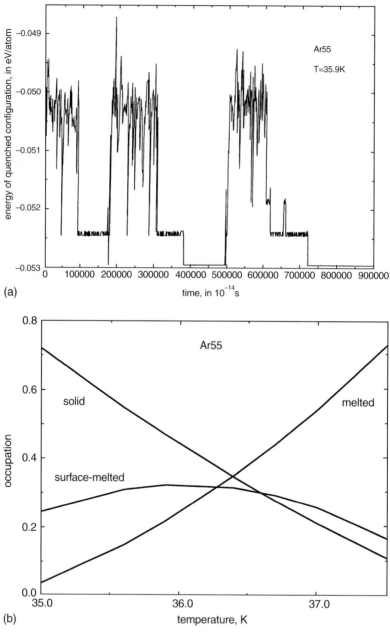

Fig. 2.4. (a) a "quench history" for an Ar_{55} cluster showing the trimodal distribution of potential energy minima, the low-energy solid, the low but intermediate surface-melted form, and the high energy liquid; (b) the fractions of the population in each of the phases, as functions of temperature, in the range in which all three forms may coexist in equilibrium

the equilibrium ratio of concentrations of clusters in two phases in dynamic equilibrium, and the other that shows the temperature ranges for one-, two- or three-phase equilibrium as a function of temperature, for a fixed pressure.

The phase diagram may be constructed from an analytic model for the partition functions of the phases in question, from simulations or from experiment. The most useful forms of phase diagram are usually those whose variables are intensive, most commonly pressure p and temperature T, but sometimes the volume V, entropy S or chemical potential μ may be useful as well. Construction of a phase diagram from an analytic model for the partition functions will be discussed below, in the context of the second kind of phase diagram. Construction of phase diagrams for clusters from experimental data is still in its infancy, so we do not discuss this in any detail. Construction of phase diagrams from simulations of cluster behavior is now a completely practical task, so we describe it now.

The simulations begin, as all molecular simulations do, with an assumption of what forces there are between the particles composing the cluster. Then one solves either a Monte Carlo computation or a set of simultaneous equations of motion, using the computational tools now well developed for these tasks [48]. The analysis of a system's behavior under conditions of constant temperature or constant pressure is straightforward via Monte Carlo algorithms. The Metropolis algorithm establishes conditions of a canonical ensemble [49]. Other variants have improved the efficiency of the search of the phase space [50–58], but the essential condition of selecting or rejecting trial points on the basis of a Boltzmann distribution of probabilities assures that the result is essentially that of an isothermal distribution.

It is not, however, so simple to carry out molecular dynamics calculations under constant-temperature or constant-pressure conditions. The first molecular dynamics simulations to be conducted with an intensive variable held constant were the constant-pressure simulations of Anderson [59]. The crux of the method was the introduction of an extra, conjugate pair of phantom variables that could act as a "ballast" degree of freedom, so that the equivalent of volume in the larger space of all the real coordinates and the new one, with all the corresponding momenta, would be constant, but in the smaller space of the real coordinates and momenta, the *pressure* would be constant, so long as all the real degrees of freedom remain in equilibrium with the phantom variables.

Nosé showed how to carry out molecular dynamics simulations so that the results would correspond to the distributions of properties of a canonical distribution; he demonstrated that previous methods based simply on rescaling velocities did not represent such ensembles. His method also required introducing a new, phantom variable so that the enlarged system maintains constant energy but the real degrees of freedom, equilibrating with the extra, new degree, maintain constant, equal *average* kinetic energies. The extra degree of freedom acts as a heat bath [60–63]. Holian, Hoover and their collaborators showed that Nosé's method could sometimes fall into a nonergodic pattern and fail to give the intended canonical distribution, but that by using two extra degrees of freedom, rather than just one, this problem could be avoided [64,65]. An alternative method for achieving a molecular dynamics simulation under constant-temperature conditions is the stochastic method of Kast et al. [66,67], used with the modification pointed out by Sholl and Fichthorn [68]. Both methods are relatively efficient; perhaps the stochastic is a little faster.

With these simulation methods, one can determine the position of the dynamic equilibrium of a system, i.e. determine the fraction of the population of an ensemble in each locally-stable phase or the fraction of time a system spends in each of those phases, for conditions of fixed temperature and pressure. In other words, one can determine the equilibrium constant at each desired temperature and pressure. For a small cluster, this may require a simulation of only a few tens of nanoseconds, but for a cluster of hundreds of particles, the time required to determine the equilibrium constant may be prohibitively long, if it is to be done by simulation of only one cluster at a time. Parallel computations, in effect combining time averaging and ensemble averaging, will certainly make such computations far more efficient.

Now we turn to the representation of the equilibrium of phases of clusters. The mode most similar to the traditional phase diagram for a pure substance of one component simply adds a new dimension. It would be possible to use the equilibrium ratio of concentrations, i.e. the equilibrium constant, K_phase = [liquid]/[solid], as the third variable. However because this quantity varies, in effect, from zero to infinity, it is far more convenient to transform to a distribution function D that we define as follows: $D(p,T) = [K_\text{phase} - 1]/[K_\text{phase} + 1]$, which is equal to [liquid] − [solid]/[liquid] + [solid] or to the difference between the amounts of liquid and solid, divided by the total amount of material. This quantity D varies from -1 to $+1$ as the fraction of liquid varies from zero to unity. The bounded variable makes visualization far easier.

With this new variable, we may construct a three-dimensional phase diagram, using pressure p and temperature T as the traditional ordinate and abscissa, with D as the other "horizontal" axis. Figure 2.5 shows two schematic examples of such diagrams, which were first presented by Cheng et al. [71]. Figure 2.5a is a schematic representation for a very large cluster or a bulk system; here the D coordinate is unnecessary because the effective values that D may take are only -1 and $+1$. In the left, low-temperature portion, D is everywhere -1, and to the right of the curve for $\Delta\mu = 0$, D is everywhere $+1$. The curved plane connecting $D = -1$ with $D = +1$ is just the traditional coexistence curve in the variables p and T spread into a surface.

The lower schematic diagram, Fig. 2.5b, is a different matter. This represents the behavior of a cluster of moderate size, one for which the unfavored phase may be observed in equilibrium, as a minority species. At the far left,

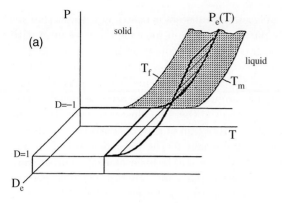

VERY LARGE SYSTEM

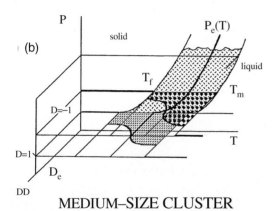

MEDIUM–SIZE CLUSTER

Fig. 2.5. Schematic representations of phase diagrams in terms of the variables pressure p, temperature T and the distribution D; (**a**) a very large system, for which the variable D is virtually redundant; (**b**) a moderate-size cluster, for which minority species appear in observable amounts even when $\Delta\mu \neq 0$

the low-temperature region, up to the beginning of the shaded regions, the system has only a single minimum in its free energy at any given p, T. so only the solid is stable and $D = -1$. At the beginning of the shaded region, at some specific temperature $T_\mathrm{f}(p)$ for each pressure, the local minimum corresponding to the liquid phase appears. This is the "freezing limit," below which only the solid is stable. The fraction of liquid increases with increasing temperature, presumably passing through the point where $\Delta\mu = 0$, up to the temperature $T_\mathrm{m}(p)$, the "melting limit" above which the free energy has again only one local minimum, now corresponding to the liquid phase. Between T_f and T_m, the two phases coexist in equilibrium. The curved surface, between the discontinuities at T_f and T_m, shows the equilibrium fraction of material in each of the phases. This is the first of the phase diagrams for small systems.

The second kind of phase diagram for clusters is based on the fact that for a system to show stability, its partition function must be an extremum with respect to the relevant parameters. If one can construct a partition function from a suitable model or from spectroscopic data, then it becomes possible to describe the phase behavior of the system in terms of the locus of points at which this stationary behavior occurs. The model used to develop this approach to phase diagrams [73,74] treated the Ar_{55} cluster by estimating the partition function, layer by layer, as a function of the concentration of defects in each layer. There was also a contribution to the total partition function from the interactions between layers. The stable form at each temperature corresponds to the extremal value of the partition function, with respect to the concentrations of defects in the core and in the surface layer. From this construction, since there are only the core and the outer shell, the partition function Z depends on only three variables (at a fixed pressure): the concentrations of defects in the core and in the surface, respectively ρ_{core} and ρ_{surf}, and on the temperature T. The locus of extrema for $Z(\rho_{core}, \rho_{surf}, T)$ or more precisely in its more convenient form, $Z(\rho_{core}, \rho_{surf}, \beta)$, where $\beta = 1/k_B T$ as usual, is a curve in the space of the three independent variables. Examples of such curves, for different choices of interparticle potentials, are shown in Fig. 2.6. The first two, Figs. 2.6a and 2.6b, were constructed with no interaction between layers; the third, Fig. 2.6c, is somewhat like real Ar_{55} but the last, Fig. 2.6d, is closest to representing the results of simulations of this cluster.

The curves are most easily interpreted by following them from the top, where the temperature is very low, along the heavy curve of the locus. The projections of the locus onto the three planes are drawn with light curves. Let us focus on Fig. 2.6d. At the lowest temperatures, there is only a single value of the partition function on the locus, which corresponds to a density of core defects near zero and a low but slowly growing value of the density of surface defects. This branch of the locus is that of the solid. The solid branch reaches a minimum and the locus then rises, corresponding to a negative heat capacity with respect to the concentration of surface defects; hence this portion, where the curve turns upward, represents an unstable branch, like the upward branch of a van der Waals isotherm, with its negative compressibility. Then the unstable branch reaches a maximum and turns downward, still at a very low value of the density of core defects but at a moderately high value of surface defects. The portion of the locus that turns downward near the left face of the cube represents the stable surface-melted form of the cluster. The curve reaches another minimum and then sweeps into the cube as the density of core defects increases, again on an *unstable* branch; at the right of the cube, in the region of $\rho_{core} = 0.9$, the curve again turns downward on the stable branch of the liquid phase.

We see that at sufficiently low temperatures, only the solid phase is stable. Then there is a very narrow temperature band in which both solid and

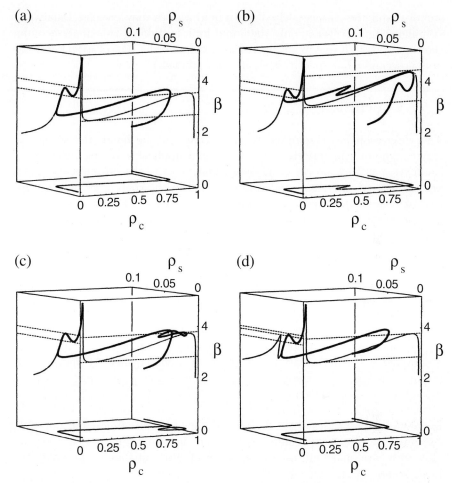

Fig. 2.6. Four examples of a second kind of phase diagram for clusters. This form displays the locus of extrema of the full partition function of the system, $Z(\rho_{\text{core}}, \rho_{\text{surf}}, \beta)$, where $\beta = 1/k_B T$, as a function of the densities of defects in the core, ρ_{core} and in the surface, ρ_{surf}, and the inverse temperature. The extrema are taken with respect to the defect densities. (a,b) are examples of possible phase diagrams for systems with van der Waals pair interactions, but no interlayer interactions between the defects. (c,d) are based on argon-like interactions with interlayer interactions between defects. Figure 6d is the figure that most accurately represents the behavior of Ar_{55} as found in molecular dynamics simulations based on Lennard-Jones interparticle interactions

surface-melted forms are stable but the liquid is not. At a slightly higher temperature, the liquid becomes stable, making it possible for all three forms, solid, surface-melted and liquid, to coexist in equilibrium over an intermediate band of temperature. At still higher temperatures, the solid loses its stability and the surface-melted and liquid phases may coexist. Finally, at still higher temperatures, only the liquid form is stable. (This analysis omits the vapor phase; it may be neglected for times of order tens or even hundreds of nanoseconds, but strictly, it should be included.)

2.5 Conclusion

This discussion has attempted to show that, on one hand, the science of thermodynamics is entirely appropriate and applicable to the behavior of small systems, and, on the other hand, that many of the concepts and interpretations that are sound and commonplace in the context of bulk matter lose or change their meaning when we deal with ensembles of very small forms of matter. Some of these are quantum properties associated with light particles being confined in volumes so small that their energy levels are forced to be relatively far from one another and the wavelike properties of the trapped masses become important. Yet the most dramatic illustration of the special thermodynamics of small systems is probably in the context of phase equilibria and phase changes. The distinction between "phase" and "component" disappears when the number of particles in an entire system is about the same as that in a component, so that the number of interactions that define a phase-like form is about like that defining the properties of a component. In bulk matter, the phase is a consequence of many, many interactions among all the particles of a macroscopic assemblage, while the properties of a component, e.g. a molecuie, come from a small number of interatomic interactions. In a cluster, only a few atoms or molecules interact to define the properties of the cluster, perhaps ten or a million–but not 10^{20}, certainly. Despite these differences between small and bulk systems, we can still, with care, make full use of the power of thermodynamics, probably the most general of all sciences.

Acknowledgements. The research was partially supported by Grant No. 3270/1999 from the Scientific Fund at the University of Sofia and Grant No. CHE-9725065 from the National Science Foundation.

References

1. L.C. Woods, *Thermodynamics of Fluid Systems*, Clarendon Press, Oxford, 1975.
2. C.E. Klots, J. Chem. Phys. **83**, 5854–5860 (1985).
3. C.C. Bradley, C.A. Sackett, J.J. Tollett and R.G. Hulet, Phys. Rev. Lett. **75**, 1687 (1995).
4. S. Grebenev, J.P. Toennies and A.F. Vilesov, Science **279**, 2083 (1998).
5. J.P. Toennies and A.F. Vilesov, Ann. Rev. Phys. Chem. **49**, 1 (1998).
6. J.L. Martins, R. Car and J. Buttet, Surf. Sci. **106**, 265 (1981).

7. W. Ekardt, Phys. Rev. B **29**, 1558 (1984).
8. W. Ekardt, Phys. Rev. B **31**, 6360 (1985).
9. D.E. Beck, Solid State Commun. **49**, 381 (1984).
10. W.D. Knight, K. Clemenger, W.A. de Heer, W.A. Saunders, M.Y. Chou and M.L. Cohen, Phys. Rev. Lett. **52**, 2141 (1984).
11. W.D. Knight, W.A. de Heer and K. Clemenger, Solid State Commun. **53**, 445 (1985).
12. W.D. Knight, W.A. de Heer, W.A. Saunders, K. Clemenger and M.L. Cohen, Chem. Phys. Lett. **134**, 1 (1987).
13. H.-P. Cheng, R.S. Berry and R.L. Whetten, Phys. Rev. B **43**, 10647 (1991).
14. R.L. Johnston, Phil. Trans. Roy. Soc. London A bf 356, 211 (1998).
15. R. Kawai and J.H. Weare, Phys. Rev. Lett. **65**, 80 (1990).
16. H. Haberland, H. Kornmeier, H. Langosch, M. Oschwald and G. Tanner, J. Chem. Soc. Faraday Trans. **86**, 2473 (1990).
17. N.F. Mott, *Metal-Insulator Transitions*, (Taylor and Francis, London, 1990).
18. M.E. Garcia, G.M. Pastor and K.H. Bennemann, Phys. Rev. Lett. **67**, 1142 (1991).
19. G.M. Pastor and K.H. Bennemann, in *Clusters of Atoms and Molecules*, H. Haberland, ed. (Springer-Verlag, Berlin, 1994), p. 86.
20. K.J. Taylor, C.L. Pettiete-Hall, O. Chesnovsky and R.E. Smalley, J. Chem. Phys. **96**, 3319 (1992).
21. X. Li, H. Wu, X.-B. Wang and L.-S. Wang, Phys. Rev. Lett. **81**, 1909 (1998).
22. J. Farges, B. Raoult and G. Torchet, J. Chem. Phys. **59**, 3454 (1973).
23. J. Farges, J. Crystal Growth **31**, 79 (1975).
24. J. Farges, M.F. DeFeraudy, B. Roualt and G. Torchet, J. Physique (Paris) **36** (C2) 13 (1975).
25. J. Farges, M.F. DeFeraudy, B. Roualt and G. Torchet, J. Physique (Paris) **38** (C2) 47 (1977).
26. A. Yokozeki and G.D. Stein, J. Appl. Phys. **49**, 2224 (1978).
27. B.G. de Boer and G.D. Stein, Surf. Sci. **106**, 84 (1981).
28. J. Farges, M.F. DeFeraudy, B. Roualt and G. Torchet, J. Chem. Phys. **78**, 5067 (1983).
29. R.K. Heenan and L.S. Bartell, J. Chem. Phys. **78**, 1270 (1983).
30. E.J. Valente and L.S. Bartell, J. Chem. Phys. **79**, 2683 (1983).
31. J. Farges, M.F. DeFeraudy, B. Roualt and G. Torchet, J. Chem. Phys. **84**, 3491 (1986).
32. L.S. Bartell, Chem. Revs. **86**, 492 (1986).
33. L.S. Bartell, L. Harsanyi and E.J. Valente, J. Phys. Chem. **93**, 6201 (1989).
34. L.S. Bartell and T.S. Dibble, J. Am. Chem. Soc. **112**, 890 (1990).
35. R.J. Mawhorter, M. Fink and J.G. Hartley, J. Chem. Phys. **83**, 4418 (1985).
36. J.G. Hartley and M. Fink, J. Chem. Phys. **87**, 5477 (1987).
37. J.G. Hartley and M. Fink, J. Chem. Phys. **89**, 6053 (1988).
38. J.G. Hartley and M. Fink, J. Chem. Phys. **89**, 6058 (1988).
39. A.L. Mackay, Acta Crystall. **15**, 916 (1962).
40. J.P.K. Doye, M.A. Miller and D.J. Wales, J. Chem. Phys. **110**, 6896 (1999).
41. J.P.K. Doye, M.A. Miller and D.J. Wales, J. Chem. Phys. **111**, 8417 (1999).
42. J. Demuynck, M.-M. Rohmer, A. Strich and A. Veillard, J. Chem. Phys. **75**, 3443 (1981).
43. V. Bonacic-Koutecky, P. Fantucci and J. Koutecky, Chem. Rev. **91**, 1035 (1991).

44. V. Bonacic-Koutecky, L. Cespiva, P. Fantucci, C. Fuchs, M.F. Guest, J. Koutecky and J. Pittner, Chem. Phys. **186**, 257 (1994).
45. V. Bonacic-Koutecky, J. Jellinek, M. Wiechert and P. Fantucci, J. Chem. Phys. **107**, 6321 (1997).
46. B.L. Holian, G.K. Straub, R.E. Swanson and D.C. Wallace, Phys. Rev. B **27**, 2873 (1983).
47. H.-P. Cheng, X. Li, R.L. Whetten and R.S. Berry, Phys. Rev. A **46**, 791 (1992).
48. *The Monte Carlo Method in Condensed Matter Physics*, K. Binder, ed., Vol. 71 in *Topics in Applied Physics* (Springer-Verlag, Berlin, 1995); D.W. Heermann, *Computer Simulation Methods in Theoretical Physics*, (Springer-Verlag, Berlin, 1990).
49. N. Metropolis, A.W. Metropolis, M.N. Rosenbluth, A.H. Teller and E. Teller, J. Chem. Phys. **21**, 1087 (1953).
50. F. Glover, ORSA J. Comput. **1**, 190 (1989).
51. F. Glover, ORSA J. Comput. **2**, 4 (1990).
52. D. Cvijovic and J. Klinowski, Science **267**, 664 (1995).
53. D.D. Frantz, D.L. Freeman and J.D. Doll, J. Chem. Phys. **93**, 2769 (1990).
54. D.D. Frantz, D.L. Freeman and J.D. Doll, J. Chem. Phys. **97**, 5713 (1992).
55. C.J. Tsai and K.D. Jordan, J. Chem. Phys. **99**, 6957 (1993).
56. I. Andricioaei and J.E. Straub, J. Chem. Phys. **53**, R3055 (1996).
57. I. Andricioaei and J.E. Straub, J. Chem. Phys. **107**, 9117 (1997).
58. F.-M. Dittes, Phys. Rev. Lett. **76**, 4651 (1996).
59. H.C. Andersen, J. Chem. Phys. **72**, 2384 (1980).
60. S. Nosé, Mol. Phys. **52**, 255 (1984)
61. S. Nosé, J. Chem. Phys. **81**, 511 (1984).
62. S. Nosé, Mol. Phys. **57**, 187 (1986).
63. S. Nosé, Prog. Theor. Phys., Suppl. 103, 1 (1991).
64. B.L. Holian, A.F. Voter and R. Ravedo, Phys. Rev. E **52**, 2338 (1995).
65. W.G. Hoover and B.L. Holian, Phys. Lett. A **211**, 253 (1996).
66. S.M. Kast, L. Nicklas, H.-J. Bär and J. Brickmann, J. Chem. Phys. **100**, 566 (1994).
67. S.M. Kast and J. Brickmann, J. Chem. Phys. **104**, 3732 (1996).
68. D.S. Sholl and K.A. Fichthorn, J. Chem. Phys. **106**, 1646 (1997).
69. F.H. Stillinger and T.A. Weber, Phys. Rev. A **25**, 978 (1982).
70. V.V. Nauchitel and A.J. Pertsin, Mol. Phys. **40**, 1341 (1980).
71. H.-P. Cheng and R.S. Berry, Phys. Rev. A **45**, 7969 (1992).
72. K. Strandburg, Rev. Mod. Phys. **60**, 161 (1988).
73. R.E. Kunz and R.S. Berry, Phys. Rev. Lett. **71**, 3987 (1993).
74. R.E. Kunz and R.S. Berry, Phys. Rev. E **49**, 1895 (1994).
75. F. Calvo and P. Labastie, Chem. Phys. Lett. **258**, 233 (1996).
76. I.L. Garzón and J. Jellinek, in *Physics and Chemistry of Finite Systems: From Clusters to Crystals*, P. Jena et al., eds. (Kluwer, Dordrecht, 1992), Vol. 1, p. 405.
77. L.S. Bartell and B.M. Powell, Molec. Phys. **75**, 689 (1992).
78. A. Proykova, R. Radev, F.-Y. Li and R.S. Berry, J. Chem. Phys. **110**, 3887 (1999).
79. A. Proykova, S. Pisov and R.S. Berry, J. Chem. Phys. (in press, 2001)
80. R.S. Berry, in *Theory of Atomic and Molecular Clusters*, J. Jellinek, ed. (Springer-Verlag, Berlin, 1999), p. 1.

3 Photoelectron Spectroscopy

G. Ganteför

3.1 Introduction

Clusters consisting of a small number of $n = 3-1000$ atoms or molecules have properties, which can be totally different from the ones of the isolated atoms and of the corresponding bulk materials. The properties of most clusters still represent a "white spot" on the map of research on nanostructures and some of these properties probably will allow for new applications. The most famous example is C_{60}, an extremely stable cluster consisting of 60 carbon atoms forming a soccer ball [1]. C_{60} is as stable as a molecule and a new bulk material consisting of bare carbon can be formed from C_{60} clusters with properties different from diamond and graphite [1]. Such very stable clusters have first been discovered in mass spectra, wherein a magic number showed up as an intense peak. For example, at the proper conditions for cluster growth C_{60} is the most abundant cluster in the mass spectrum of bare carbon clusters [2]. Analogously, in the mass spectra of clusters of simple metals like e.g. K_n clusters, magic numbers corresponding to $n_e = 8, 18, 20$ and 40 valence electrons show up as outstanding lines or steps [3]. Based on these data models like the Jellium model for the metal clusters [4] have been developed to explain the high stabilities of these magic number clusters. However, the relative abundance of a cluster of a certain size in the mass spectrum contains no structural information and a reliable test of any model predicting the electronic and geometric structure of a cluster needs support by additional spectroscopic data. With a diameter of typically 1 nanometer, these clusters are so small that common microscope techniques like electron microscopy and scanning tunneling microscopy fail for most cases. Therefore, spectroscopy of clusters in the gas phase is employed to gain structural information.

Any spectroscopy of mass-selected clusters is not easy. Common cluster sources generate a broad size distribution of clusters and spectroscopic data have somehow to be assigned to a certain cluster size. One successful solution of this problem is the spectroscopy of cluster ions instead of neutrals. Therefore, in the early days of cluster research several new spectroscopic techniques dealing with ions have been developed, where the cluster ions are mass-selected prior to the experiment. The problem of experiments with ions is the low intensity compared to neutrals. Typically, an ion beam has a target density of about 6 orders of magnitude lower than the density of neutrals in, e.g., supersonic beams. Therefore, the spectroscopic techniques have to be extremely sensitive. To deal with this problem, in photoelectron spectroscopy a laser is used as the light source instead of a standard gas discharge lamp, gaining about 4 orders of magnitude of intensity. Still, additional tricks are necessary to improve signal to noise ratio compared to standard photoelectron spectroscopy experiments on neutral atoms, molecules and surfaces [5,6].

There are many different experimental techniques for spectroscopy of mass-selected cluster ions and here we will focus just on one: photoelectron spectroscopy of mass-selected anions (PES$^-$) [7–12]. A single photon generated by a laser interacts with a negatively charged cluster and detaches a photoelectron. The kinetic energy of the photoelectrons is determined. The method cannot be applied to positive ions, because the minimum energy which is required to remove an electron from a positive ion is considerably higher than to remove it from an anion. Commercial UV-lasers generate light with photon energies up to 6.4 eV or 7.9 eV, which is not sufficient for most positive cluster ions.

To illustrate the kind of information to be gained using PES$^-$ it can be compared to a complementary spectroscopic technique: ion mobility [13–15]. Its basis is the measurement of the friction of a moving cluster ion dragged by an electric field through an inert gas (Helium). The friction coefficient is higher, if the cluster has an open structure like, e.g., a chain or ring structure; and smaller, if the cluster is compact. In the case of C_{60}, ion

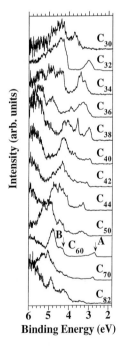

Fig. 3.1. Photoelectron spectra of fullerenes. The photon energy is $h\nu = 6.4$ eV. The distance between the first peak at lowest binding energy and the next feature at higher binding energy corresponds to the HOMO-LUMO gap (see next section), which is extremely large for the magic cluster C_{60} (indicated by arrows, reproduced from [20])

mobility measurements support the assumed structure of a hollow sphere for all clusters with sizes larger than $n > 30$. However, the experiment indicates no special behavior of C_{60} and its mobility is just in between the one of C_{58} and C_{62} [16]. The high stability of C_{60} is due to its detailed geometric structure, which cannot be resolved using ion mobility and which is directly related to its electronic structure [1]. However, the electronic structure can be examined via PES$^-$ and the photoelectron spectrum of the C_{60}^- anion [17–20] is totally different from the ones of all other carbon cluster anions (Fig. 3.1).

The above example of PES$^-$ spectra demonstrates the power of the technique, which can be applied to study all kinds of different bare and reacted clusters with unknown electronic and geometric structure. In the following Sect. 3.2, an introduction into the physics of photoelectron spectroscopy is given. In Sect. 3.3 a typical experimental set up is described and in Sect. 3.4 several examples of experimental results with the corresponding interpretation are discussed.

3.2 Physics of Photoelectron Spectroscopy

In general, photoelectron spectra contain information about the electronic structure, i.e. the energetic ordering of the bonding, non-bonding and anti-bonding single particle orbitals. There are two pictures, which are used for interpretation of photoelectron spectra. The most simple one, which is less accurate, is the single particle picture (Fig. 3.2). It is assumed, that the photon interacts with a single electron in the cluster anion, which is detached, if the photon energy $h\nu$ is higher than the binding energy (BE) of the electron. The kinetic energy E_{kin} of the outgoing electron is:

$$E_{kin} = h\nu - BE .$$

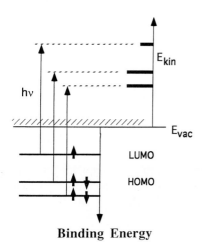

Fig. 3.2. Single particle picture of photoemission. By the interaction with a photon with energy $h\nu$ an electron is ejected with kinetic energy E_{kin} corresponding to the difference of $h\nu$ and the binding energy of the single particle orbital. All energies refer to the vacuum energy E_{vac} corresponding to zero binding energy. If the neutral cluster has a closed electronic shell, the HOMO (highest occupied molecular orbital) is completely filled. In that case, the additional electron of the anion occupies the LUMO (lowest unoccupied molecular orbital) of the neutral

In each event only a single photon interacts with a single cluster (at low photon flux). A photoelectron spectrum is the kinetic energy distribution of the electrons and corresponds to an average over many such processes. If there are electrons bound in different single particle orbitals, in the electron spectrum various peaks with different kinetic energies will appear. The relative intensities of the peaks reflect the probability of a certain process and depend on the number of electrons occupying the orbital, the symmetry of the orbital and the photon energy [5]. According to the above considerations, a spectrum is a picture of the occupied single particle orbitals in the cluster up to a maximum BE, which corresponds to the difference of the photon energy and the electron affinity (EA). A higher photon energy and lower EA gives a deeper insight into the valence orbital structure.

In principle, the single particle picture holds for neutral and for negatively charged particles. However, there is one important advantage for anions. If the neutral particle has a closed electronic shell, i.e. the highest occupied molecular orbital (HOMO) is completely filled, the additional electron of the corresponding anion must occupy the lowest unoccupied molecular orbital (LUMO). That is, for closed shell species (this notation always refers to the neutral particle) in the PES a peak at high kinetic energy appears due to the detachment of the weakly bound additional electron in the LUMO (Fig. 3.2). The difference in kinetic energy between this peak and the next feature at higher BE corresponding to the detachment from the HOMO gives the HOMO-LUMO gap of the neutral. This energy is an important parameter of a cluster closely related to its stability [20]. For example, in Fig. 3.1 such a peak at low BE with a large gap to the next feature appears for C_{60} (marked by arrows). Accordingly, directly from the PES$^-$ a high stability of C_{60} can be predicted. In photoelectron spectroscopy of neutral particles the HOMO-LUMO gap cannot be determined.

The overly simple version of the single particle picture described above must be improved by the following effects to gain at least crudely quantitative results [21]:

– *relaxation*
The detachment of an electron increases the charge state and, as a result, the BEs of all other single particle orbitals increase. This relaxation energy is transferred to the outgoing electron increasing its kinetic energy.

– *multiplet splitting*
In the final state (the neutral state) the spins and angular moments of the remaining electrons can be combined to different total spins and angular moments with slightly different total energies. Depending on this energy of the final state the kinetic energy of the outgoing electron is different. Accordingly, the detachment of an electron from a certain single particle orbital might give rise to more than one peak in the PES$^-$ (multiplet).

– *"shake up" process*
The detachment might cause another bound electron to be excited into a higher bound single particle orbital. This excitation energy lacks the kinetic energy of the outgoing electron. In this case, the photon interacts with two electrons.

– *configuration interaction*
If especially from an orbital at relatively high BE an electron is detached, the remaining orbitals might undergo a strong disturbance and there might only be a limited similarity of shape and energetic ordering of the final state orbitals with the ones of the initial state. In this case, the final state can no longer be described by the initial state configuration with one electron missing, but must be described by a linear combination a various initial state configurations. This corresponds to the break down of the single particle picture. In many cases, the single particle picture cannot be applied to assign peaks at relatively high binding energies [21].

A more accurate model of the photoemission process is the description of the photoemission process as a transition from an initial all-electron state X^- with certain total energy, spin and angular momentum into all-electron final states X, A, B ... (Fig. 3.3). While in the single particle picture the BEs of the valence orbitals are considered, here the total energy of the state formed by all electrons counts. In the spectrum the feature at lowest BE corresponds to the transition from the anion electronic ground state X^- (assuming "cold" anions generated by the source) to the electronic ground state X of the neutral cluster. The next peak at higher binding energy is assigned to the transition into the first excited state A of the neutral. That is, the photoelectron spectrum of the anion is a map of the electronic states of the neutral cluster.

The dipole selection rules have to be applied in order to fully interpret the process. The outgoing electron can carry different angular moments and, therefore, usually transitions into almost all neutral states are allowed [5]. That is, more neutral states than in conventional photoabsorption spectroscopy (e.g., resonant two photon photoionization spectroscopy of neutrals) are visible using PES^- [22]. There is only one selection rule concerning the spin: the outgoing electron carries a spin of 1/2 and, accordingly, the total spin of the initial and final state can differ by 1/2 only (for LS-coupling). For example, if the anion ground state is a singlet state, only transitions into neutral doublet states are allowed.

Within this picture the width of a peak and its vibrational fine structure – if resolved – can be understood, too [7,23,24]. The initial state might be the electronic ground state of the anion. The anion has also vibrational degrees of freedom and, if the anion is "cold", it is in the vibrational ground state $\nu = 0$. The final states also have vibrational degrees of freedom and not only transitions into the vibrational ground state, but also into excited vibrational states $\nu' = 1, 2, \ldots$ might be allowed (Fig. 3.3). That is, for a certain

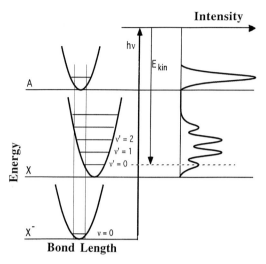

Fig. 3.3. All – electron picture of the photoemission process. The process corresponds to a transition from the anion electronic ground state X⁻ into various electronic states X, A, B ... of the neutral. Each state has vibrational degrees of freedom and a certain equilibrium geometry (= bond length in case of a dimer). If the geometries of initial and final state are similar, only vibrational transitions between similar quantum numbers (e.g., 0 ⟶ 0 transitions) are observed and sharp peaks appear in the spectrum. A large change in geometry results in the appearance of a broad peak with an extended vibrational fine structure (Franck-Condon principle). The fine structure corresponds to vibrational excitations of the neutral cluster

electronic transition several vibrational transitions are possible with slightly different energies resulting in a broadening of the corresponding peak in the PES⁻. If the energy resolution of the electron spectrometer is sufficient, the single vibrational transitions might be resolved [23,24]. The intensities of certain vibrational transitions are governed by the Franck-Condon transition probabilities and depend on the difference of the equilibrium geometries of the two states. If both geometries are identical, only 0 ⟶ 0 transitions are allowed (for "cold" anions) and the peaks in the PES⁻ are sharp. Broad peaks in the PES⁻ indicate a large difference of the equilibrium geometries of the anion and the neutral. There is rotational fine structure in the PES, too. However, it is usually not observed because of the relatively low experimental resolution of PES and will be neglected for the forthcoming discussion.

As an example the PES⁻ of Cu_1^- displayed in Fig. 3.4 will be assigned using the two pictures. The spectrum has been recorded with a photon energy of $h\nu = 6.4\,\mathrm{eV}$. The spectrum is plotted versus the BE, which is given by $\mathrm{BE} = E_{\mathrm{kin}} - h\nu$. This has the advantage that the peak positions do not depend on the photon energy. The configuration of the valence electrons of the Cu_1^- anion is $3d^{10}4s^2$. Within the single particle picture two peaks are

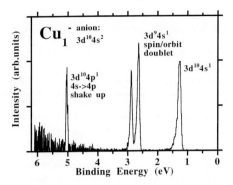

Fig. 3.4. Photoelectron spectrum of the Cu_1^- anion recorded with a photon energy of 6.4 eV. The intensity of the detected photoelectrons is plotted versus the binding energy (= photon energy minus kinetic energy). The widths of the peaks correspond to the experimental resolution. For an assignment of the features see text

expected due to emission from the two occupied orbitals 3d and 4s. Indeed, the single peak at ≈ 1.2 eV BE can be assigned to emission from the 4s orbital and the doublet at ≈ 2.8 eV to emission from the 3d orbital. Since the remaining nine 3d electrons can combine to two different spins and angular moments, two peaks are observed and the splitting is caused by the spin-orbit interaction in the final state [25]. Because of its zero angular moment, there is no such splitting for the 4s peak. The peak at ≈ 5 eV BE is assigned to a shake up process. While one of the 4s electrons is detached, the other one is excited into the unoccupied 4p orbital. This energy lacks the outgoing electron and gives rise to a peak at low kinetic energy (= high binding energy).

In the all-electron picture, the peak at lowest BE is assigned to the transition from the anion electronic ground state $^1S_0$ to the neutral ground state $^1S_{1/2}$ [7,25]. The doublet is assigned to transitions into the neutral excited states $^2D_{5/2}$ and $^2D_{3/2}$ and the peak at 5 eV BE is the transition into the $^2P_{1/2}$ state of neutral Cu [25]. The BE of the first feature corresponds to the electron affinity of Cu (1.23 eV [26]) and the difference in BE between this peak and the next one at 2.6 eV is the energy of the first excited state of the neutral Cu atom [25]. This demonstrates that the PES$^-$ is indeed a map of the electronic states of the neutral. An analysis of the configurations of the all-electrons states shows that the two pictures are complementary [25].

Within the all-electron picture, the energy difference between the first two peaks in the spectrum is the excitation energy of the neutral first excited state. This corresponds to the HOMO-LUMO gap of molecules and clusters and to the band gap of semiconductors. We conclude, that there are three main advantages of photoelectron spectroscopy of negatively charged particles:

- easy mass separation
- direct determination of the HOMO-LUMO gaps of closed shell species
- mapping of the neutral electronic states.

3.3 Experimental Set Up

An experimental setup to record photoelectron spectra of mass-selected cluster anions consists of 4 components: (i) cluster source, (ii) mass-spectrometer, (iii) laser and (iv) electron spectrometer. The various different experimental setups can be divided into two types of apparatus: continuously and pulsed operated experiments. The pioneering experiments from W.C. Lineberger's and K. Bowen's groups [7–9,23] were operated with a continuous source (oven source, hollow discharge source), a quadrupole mass spectrometer, a continuous wave laser and a hemispherical analyser for electron spectroscopy. Such a setup has certain advantages: a relatively high energy resolution ($< 3\,\text{meV}$) and the possibility of angle-resolved PES^-. One disadvantage is the limited photon energy of continuous lasers (e.g., $h\nu = 3.53\,\text{eV}$ in [23]). Pulsed PES^- experiments use a pulsed cluster source (laser vaporization [10], pulsed arc cluster ion source [24]), a time-of-flight mass spectrometer, a pulsed UV-laser (Nd:YAG laser, excimer laser), and a time-of-flight electron spectrometer [10–12,17–22,24,27–35]. One advantage of this setup is the relatively high photon energy provided by excimer lasers (ArF: 6.4 eV and F_2: 7.9 eV). If a "magnetic bottle" type ([36], see below) time-of-flight electron spectrometer is used, the time for recording a spectrum with reasonable signal to noise ratio can be as short as 1 min [24]. The pulsed version of a PES^- experiment can also be combined with a femtosecond laser system to study fast dynamical processes in clusters [37–41]. In the following, one example of a pulsed experimental setup [42] will be described in more detail (Fig. 3.5).

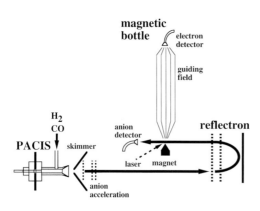

Fig. 3.5. Experimental set up of a pulsed PES^- experiment [42]. The cluster anions are directly produced in a pulsed arc cluster ion source (PACIS). Reacted cluster anions can be generated by introducing gases like CO or H_2 into the extender. The anions are mass-separated by a reflectron time-of-flight mass spectrometer. A selected bunch is irradiated by a UV-laser pulse and the kinetic energy of the detached photoelectrons is measured with a "magnetic-bottle" type time-of-flight electron spectrometer

3.3.1 Cluster Source

The cluster anions are generated using a pulsed arc cluster ion source (PACIS [43–46]. This source is basically similar to the laser vaporization source [47,48], but the bulk material is vaporized by an electric arc instead of a laser pulse. The pulsed electric arc burns within the center of ceramic cube. The cube has two channels perpendicularly intersecting each other in the center. The electrodes are inserted from opposite sides into the larger bore and face each other at the center of the cube leaving a gap. A carrier gas pulse (usually Helium) flows through the second channel and the gap between the electrodes. The carrier gas leaves the cube on the opposite side streaming into the extender. Simultaneously with the gas pulse the electric arc is ignited resulting in vaporization of material from the cathode. The metal vapor and the helium mixes within the discharge and the resulting plasma is flushed into the extender. Finally, the clusters embedded into the carrier gas leave the source through a conical nozzle forming a supersonic expansion. The source directly produces negative ions. Reacted clusters can be produced by introducing reactive gases like CO or H_2 into the extender.

3.3.2 Time-of-Flight Mass Spectrometer

After passing a skimmer the anions embedded in the supersonic beam are accelerated up to a kinetic energy of about 1000 eV by a pulsed electric field. Because of their different mass the anions have different velocities and the anion cloud separates into a chain of bunches with well defined masses. To improve the mass resolution, the anion beam is guided into a reflectron, which operates like an electrostatic mirror [49]. The reflectron compensates for small differences in the kinetic energies of the anions. The faster anions dive deeper into the reflectron field and, therefore, need a slightly longer time for reflection. At a proper choice of distances and field strengths this compensates the shorter time-of-flight of the faster anions and the mass resolution can be improved by several orders of magnitude with respect to a linear Wiley-McLaren type time-of-flight spectrometer [50]. This is especially important for the spectroscopy of reacted clusters like $Ti_n H_m^-$ clusters (see below). Because of the need of extremely high anion intensities the mass resolution of the discussed setup is limited to $m/\Delta m = 400$, although in principle with such types of mass spectrometers a resolution of > 35000 can be achieved [51].

3.3.3 Laser

The anion beam is guided into the center region of the electron spectrometer and a selected bunch is irradiated by a UV-laser pulse. Most common lasers are Nd:YAG lasers with photon energies of $h\nu = 2.33$ eV, 3.49 eV and 4.66 eV and excimer lasers with photon energies of $h\nu = 6.4$ eV and 7.9 eV. The mass is selected by tuning the delay between the acceleration of the anions and the firing of the laser. The photon flux should be kept as low as possible to minimize multiphoton processes. At high flux, a certain cluster anion might first absorb a photon and thermalize its energy. If a second photon detaches an electron from this hot cluster, a broad spectrum will appear. The first photon might also cause fragmentation and a fragment spectrum is superimposed on the desired 1-photon spectrum. If a "magnetic-bottle" type spectrometer (see below) is used, spectra can be recorded with laser pulse energies smaller than 10 µJ avoiding any such processes.

At photon energies similar or higher than the work function of the surfaces in the electron spectrometer (typically graphite with a work function of about 5 eV) many background electrons are emitted from the surfaces by stray light of the laser. Therefore, a careful collimation of the laser beam is necessary.

3.3.4 Photoelectron Spectrometer

For a pulsed experimental set up the kinetic energy of the detached electrons can be determined with the time-of-flight method. The most simplest version is a drift tube with an electron detector at the end (typically 1 m long) [52]. The longer the tube the higher is the energy resolution. However, the electron intensity decreases with increasing distance from the center region. With such a setup an energy resolution of better than 10 meV can be achieved but a relatively high photon flux is needed to gain intensity [53].

Magnetic fields can be used to guide almost all electrons from the center region to the electron detector [36]. Such a spectrometer is called "magnetic-bottle" [10,24] because of the special shape of the magnetic fields. The center region, where the anion and the laser beam intersect, is located in a strong divergent magnetic field (~ 0.1 T/mm). In such a field the electrons follow the gradient towards the weak field region (~ 0.001 T). Within the weak homogenous magnetic field the electrons are guided through a distance of about 1–2 m to the electron detector. With this setup, a gain in electron intensity by about 2 orders of magnitude is achieved allowing for experiments with extremely low laser intensity.

There is a disadvantage: The velocity of the electrons v_e measured by the spectrometer is the vector sum of the center-of-mass velocity of the electrons v_{cm} and the velocity of the anions v_a

$$\boldsymbol{v}_e = \boldsymbol{v}_{cm} + \boldsymbol{v}_a .$$

Since almost all electrons emitted in various directions with respect to the anion velocity are collected, different v_e's are measured depending on the angle between the two velocity vectors. The maximum "Doppler" broadening of v_e is $2 \times v_a$. The resulting uncertainty of the kinetic energy of the electrons dE is (m_e = electron mass):

$$dE \sim 2 m_e \times v_{cm} \times v_a .$$

For example, photoelectrons with a kinetic energy of 1 eV ($v_e = 590 \times 10^3$ m/s) emitted from clusters with a mass of 500 amu and a kinetic energy of 1000 eV ($v_a = 20 \times 10^3$ m/s) exhibit a Doppler broadening of 130 meV [10]. This limitation can be overcome by a deceleration of the anions right in the center region [24]. With this deceleration technique the kinetic energy of the anions can be reduced to about 10–50 eV narrowing the Doppler broadening down to <10 meV (Fig. 3.6). With this technique for many clusters vibrational fine structure can be resolved [24].

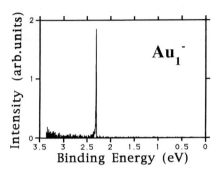

Fig. 3.6. Photoelectron spectrum of Au_1^- recorded with a photon energy of $h\nu = 3.49$ eV. The FWHM of the single feature is 10 meV

Nowadays, the pulsed setup using a "magnetic bottle" spectrometer is one of the most common used for photoelectron spectroscopy of mass-selected cluster anions [10,11,17–22,24,27–35,37–42].

3.4 Results

There are many PES$^-$ experiments of the continuous and pulsed type running in many countries all over the world and there is a wealth of photoelectron data. Only an arbitrary selection of a few examples can be discussed here. First, some spectra of clusters of simple metals will be displayed revealing the electronic shell structure predicted by the Jellium model. Second, with decreasing cluster size the band width of the conduction and the valence bands in semiconductor nanoparticles decrease and, as a consequence, the band gap is expected to increase. Some results on "semiconducting" clusters will be presented with surprising results. The third example will shed some light on the fundamental difference of chemical properties of clusters and of bulk surfaces. Finally, some data on the ultrafast dynamical processes in clusters using femtosecond lasers will be presented.

3.4.1 Example: Electronic Shells in Clusters of Simple Metals

In the mass spectra of clusters of monovalent (= simple) metals like Na, K, Cu, Ag and Au the observed magic numbers are different for different charge states and appear at certain numbers of delocalized electrons n_e in the cluster [54,55]. These magic numbers are $n_e = 2, 8, 18, 20, 34, 40, \ldots$; i.e., especially abundant clusters are e.g. Na_7^-, Na_8 and Na_9^+. This observation can be explained with the Jellium model [4,54]. It assumes a spherical shape of the metal cluster and that the delocalized s-electrons can move freely within the cluster. This corresponds to the quantum mechanical problem of electrons in a spherical potential similar to protons and neutrons in atomic nucleii. The single particle eigenstates have well defined angular momenta with a certain degree of degeneracy. The lowest single particle orbitals are: $1s^2$, $1p^6$, $1d^{10}$, $2s^2$, $1f^{14}$, $2p^6$, etc. In parenthesis the number of electrons which can be accommodated in the shell is given. The sum of these numbers corresponds to the observed magic numbers of n_e. If a shell is completely filled, the cluster has a closed electronic shell like a rare gas atom in the periodic table and a high ionization potential or electron affinity. A cluster with one electron more has to accommodate this electron in the next shell and is hence similar to an alkali atom. It has a low ionization potential or electron affinity.

The goal to reveal the electronic shells in the cluster is an ideal task of photoelectron spectroscopy. Within the single particle picture for, e.g., Na_7^-, only two peaks are expected due to the emission from the 1s and from the 1p shell. The pioneering PES$^-$ experiments on clusters of simple metals have been conducted by the groups of W.C. Lineberger (Cu, Ag, Au) [2,23] and K. Bowen (Na, K, Cs) [9,54]. At higher photon energy but with moderate energy resolution Smalley's group has presented data on clusters of the coinage metals [56,57]. Here, we will discuss just three selected spectra (from Refs. [21,58–61]) of Na_7^-, Ag_7^- and Ag_8^- to illustrate the main ideas (Fig. 3.7).

The spectrum of Na_7^- shows a narrow peak at 2.3 eV BE and a broad feature centered at 1.5 eV BE [61]. The broad feature exhibits a splitting into three peaks. We assign the spectrum using the single particle picture. In the shell model, the configuration of the Na_7^- cluster is $1s^2 1p^6$. Each Na atom contributes one delocalized s-electron to the cluster (plus the additional electron of the anion). The peak at high BE is assigned to emission from the 1s shell and the broad feature to emission from the 1p shell. The splitting into three peaks can be explained by the non-spherical symmetry of this cluster lifting the degeneracy of the 1p suborbitals (noted $1p_{1,2,3}$). The spectrum of Na_7^- demonstrates the power of PES$^-$, revealing the electronic structure of the valence orbitals in clusters. It should be noted, that for this cluster with an relatively low electron affinity, even at the low photon energy used in the experiment ($h\nu = 3.49$ eV) the complete valence "band structure" can be analysed.

In Fig. 3.7b,c the effect of an additional electron added to the closed shell is demonstrated for Ag_n^- clusters [21]. As in the case of Na, each Ag atom contributes one s-electron to the lake of electrons delocalized in the cluster.

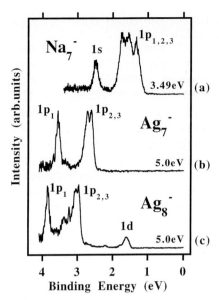

Fig. 3.7. Photoelectron spectra of Na_7^- (**a**), Ag_7^- (**b**) and Ag_8^- (**c**), recorded with photon energies of $h\nu = 3.49\,eV$ (**a**) and $h\nu = 5.0\,eV$ (**b,c**), respectively. The assignments of the peaks refer to the electronic shells predicted by the Jellium model (see text). In a non-spherical potential the threefold degeneracy of the 1p orbital is lifted and the suborbitals noted $1p_{1,2,3}$ appear at different binding energies

The 4d-orbitals are located at higher BE ($> 5\,eV$) and can be neglected here. Compared to the alkali metals, the BEs of all valence electrons are higher in Ag and the emission from the 1s shell cannot be observed with the relatively low photon energy of $h\nu = 5\,eV$. The three peaks observed in Fig. 3.5b can be assigned to emission from the $1p_{1,2,3}$ orbitals analogous to the case of Na_7^- with lifted degeneracy due to imperfect spherical symmetry of the Ag_7^- cluster. In the spectrum of Ag_8^- (Fig. 3.7c) a similar group of three peaks is observed with a slight shift towards higher BE. This is to be expected for a "shell" model: with increasing cluster size the shell moves "inwards". The 9th electron must occupy the next shell: the 1d shell. Indeed, at low BE a small peak (marked 1d) appears. The spectra demonstrate that the clusters Ag_7^- and Ag_8^- exhibit a certain similarity to a rare gas atom with a closed electronic shell and an alkali atom with a single weakly bound electron, respectively.

The three spectra discussed above also can be quantitatively analysed by comparison with corresponding calculations [62]. These calculations support the qualitative interpretation given above and also can explain some fine structure in the spectra not discussed here.

3.4.2 Example: The Size Dependence of the Band Gap

In the transition from the atom to the bulk the sharp atomic levels broaden with increasing cluster size and finally form broad bands. In all cases – atoms, molecules, clusters, nanoparticles, bulk materials – the "bands" consist of a number of discrete single particle orbitals. The only difference is, that this number increases linearly with the number of atoms in the particle. For example, a Na_2 dimer has two single particle orbitals forming the valence band (the bonding and anti-bonding 3s-derived orbitals) and bulk Na has 10^{23} single particle orbitals forming the valence band (= conduction band in the case of metals). With a bandwidth of roughly 10 eV this corresponds to a density of states of about 10^{22} states/eV and an average distance of 10^{-22} eV between neighboring orbitals. Even a piece of bulk Na with 10^{23} atoms has a band gap, but it is extremely small compared to the thermal energy at room temperature and it behaves like a metal. Small metal clusters have a low density of states and, e.g., the "band gap" (= HOMO-LUMO gap) of Ag_8 is about 1.3 eV (Fig. 3.7c) and from this point of view the cluster can be considered to be semiconducting. With increasing cluster size this "band gap" of the metal clusters decreases and the cluster become "metallic".

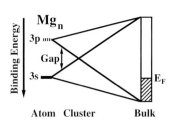

Fig. 3.8. Schematic of the development of the 3s- and 3p-derived density of states in Mg clusters with increasing size. In bulk Mg, the 3s- and 3p-derived orbitals are hybridized and 25% of the band is occupied (2 electrons and 4 single particle orbitals per atom). E_F is the Fermi energy. In small clusters, there is a gap between the completely occupied 3s- and the empty 3p-derived orbitals and these clusters can be considered to be semiconducting

For the two-valent metals like Mg with the atomic configuration $3s^2$ the situation is more complex. In bulk Mg, the valence band is formed by contributions from the atomic 3s and 3p orbitals. Since in the atom the 3s is located at higher BE than the 3p orbital, in small clusters the two bands (= the two groups of 3s- or 3p- derived orbitals) are still separated. The 3s-band is completely filled and the 3p-band empty like the conduction and valence bands in a bulk semiconductor (Fig. 3.8). Small Mg clusters are expected to have a band gap and with increasing cluster size both, the 3s and 3p band broaden and finally overlap. The gap vanishes and the cluster becomes metallic. The size-dependence of the "band gap" in such clusters can be monitored using PES$^-$ and recently, such experiments have successfully been conducted on Hg clusters in O. Cheshnovsky's group [29]. The Hg atom with the configuration $5d^{10}6s^2$ is similar to the Mg atom, if the completely filled 5d orbital is considered as a closed "inner" shell. The series of PES$^-$ spectra of

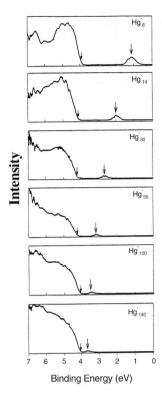

Fig. 3.9. Photoelectron spectra of some selected Hg_n^- clusters recorded with a photon energy of $h\nu = 7.9$ eV (reproduced from [29]). The small peak at low BE (marked with a large arrow) is assigned to detachment of the additional electron occupying the empty 6p band, while at high BE the onset of the emission from the fully occupied 6s band is visible (marked with a small arrow)

Hg_n^- clusters presented in Fig. 3.9 demonstrate the closing of the gap between the 6s and 6p "bands" occurring at a cluster size of about $n = 400$. The small peak at low BE (marked with the large arrow) is assigned to detachment of the additional electron occupying the empty 6p band, while at high BE the onset of the emission from the fully occupied 6s band is visible (marked with a small arrow).

What happens in case of clusters of a real semiconductor like Si? According to the simple consideration above the "band gap" is expected to increase. In bulk Si the gap is 1.1 eV. Fig. 3.10 displays a comparison of results of calculations with experimental photoelectron spectra of Si_n^- clusters with $n = 8-20$ [63]. From the comparison of theory and experiment, the spectral features can be unambiguously be assigned and the gaps can be determined. The size dependence of the gaps is displayed in Fig. 3.11 for $n = 3-20$ calculated for clusters in their neutral equilibrium geometry. In many cases the gaps are larger by a factor of 2 than the band gap of bulk Si. However, the sizes of the gaps vary irregularly with cluster size and no clear size dependence can be extracted.

The main difference between Si on one hand and Mg or Hg on the other hand is the type of bonding. The covalently bound Si clusters have a well-

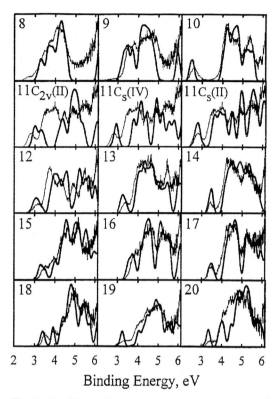

Fig. 3.10. Photoelectron spectra of Si_n^- anions ($n = 8-20$), measurements (*thin lines*), and theoretical simulations (*thick lines*). For $n = 11$, the PES simulated for several different low energy isomers are shown. The isomer $C_s(II)$ gives the best agreement with the experimental data (reproduced from [63])

defined geometry because the bonds are highly directional. Dangling bonds at the surface of a Si cluster cost an enormous amount of energy (roughly 10 eV) and the cluster rearranges its geometry to minimize the number of dangling bonds. A similar reconstruction occurs at bulk Si surfaces. The reconstruction depends on the number of atoms in the cluster and, therefore, the geometric and electronic structures of the clusters vary irregularly. In

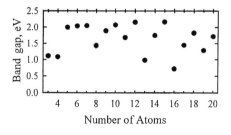

Fig. 3.11. Band gaps calculated for the lowest energy isomers of neutral Si_n clusters (reproduced from [63])

all cases, the geometric structure of the Si clusters differs totally from the "diamond" structure of bulk Si. For the clusters with $n = 9-18$ the geometric structure bases on a tri-capped trigonal prism (Fig. 3.12) [63,64].

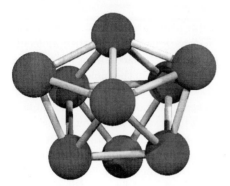

Fig. 3.12. Tricapped trigonal prism (TTP) structure of the Si_9 cluster. The geometries of the larger clusters with up to $n = 20$ atom base on TTP subunits [63,64]

3.4.3 Example: Chemical reactivity and electronic structure

Small metal particles have exceptional chemical properties. In heterogeneous catalysis such particles supported on metal oxide substrates enhance the yield of certain products of chemical reactions. For commercial catalysts, the particle size, shape and adsorption site on the substrate are relatively undefined and still, the average catalytic activity is high. That seems to indicate, that exact size and detailed geometric structure of the supported clusters are less important. However, some experiments on the chemical reactivity of unsupported mass-selected clusters show a dramatic size dependence and the rates vary in some cases with each additional atom by orders of magnitude [65–72]. In addition, experiments and calculations on supported clusters show the importance of the detailed size and geometric structure and of the interaction with the substrate [73,74].

There seem to exist two effects: (i) a general difference in the chemical properties of clusters compared to bulk surfaces, which varies only smoothly with cluster size and (ii) a strong variation of the chemical properties of each individual cluster depending on its detailed size and structure. In the following we will discuss an idea about the origin of a possible general difference between the chemical properties of clusters and bulk surfaces.

One general difference, which is independent on the detailed shape of the cluster, is the size dependence of the ionization potential ($\iff$ work function) [75]. The ionization potentials of the clusters decrease with increasing size approaching the bulk work function. This is related to the average coordination number and a result of an ineffective screening of the hole created by the emission of the photoelectron. In the bulk, a hole is immediately and effectively screened by the valence electrons from the next-neighbor atoms. This effective screening reduces the energy necessary to remove the electron and, accordingly, ionization of an isolated atom takes more energy than ionization of a bulk atom.

As a consequence, electron transfer from a metal cluster into an empty and, possibly, anti-bonding orbital (charge donation) of an adsorbed molecule (e.g., H_2, CO) might be small compared to the bulk surface. In case of a strong charge transfer into an anti-bonding orbital the bond in the molecule is broken and the molecule is dissociatively bound. In case of a cluster, the reduced charge transfer might not be sufficient to break the bond and the molecule is bound intact. Such a dependence on the work function is well known in surface chemistry for the case of CO chemisorption on transition metal surfaces [76]. CO is bound as a molecule to Cu, Ni or Pd surfaces, while it dissociates on surfaces of the "early" transition metals (i.e., the ones on the left side of the periodic table) like Sc or Ti. This effect is explained by the differences in the work functions between the "early" and the "late" transition metals. For a low work function (Sc, Ti) the d-electrons can be removed more easily and charge donation from the metal d-band into the anti-bonding $2\pi^*$ orbital of CO is increased. The bond is broken and CO is bound dissociatively. For a higher work function there is insufficient charge transfer and CO is bound molecularly. An analogous transition should take place for clusters with decreasing size and increasing ionization potential.

We studied the chemisorption of hydrogen on small Ti clusters. H_2 chemisorption on Ti bulk surfaces is dissociative [77]. For a single Ti atom, H_2 chemisorption should be non-dissociative because of energetic considerations: the binding energy of the TiH diatomic molecule is 1.96 eV [78], while the H_2 molecule is bound by 4.48 eV [79]. As long as the binding energy of H_2 is larger than twice that of a single Ti-H bond, chemisorption will be non-dissociative. In contrast, a single hydrogen atom is bound by greater than 3.0 eV to a Ti surface, so for bulk surfaces dissociative chemisorption is the lower energy channel with a total gain in binding energy of 2×3.0 eV -4.48 eV or greater than about 1.5 eV [80]. Accordingly, a size-dependent transition from molecular to dissociative H chemisorption must occur for Ti_n clusters.

In the mass spectra, features assigned to a particular cluster consist of several narrow peaks reflecting the isotope distribution of Ti. Each $Ti_nH_m^-$ cluster gives rise to a single intense peak accompanied by two weaker lines on both sides. As an example, Fig. 3.13a displays two features in the mass spectrum of the $Ti_nH_m^-$ clusters, which are assigned to $Ti_2H_m^-$ and $Ti_5H_m^-$ [81]. The intensity distribution of the feature assigned to $Ti_2H_m^-$ is a superposition of contributions from $Ti_2H_4^-$ and $Ti_2H_6^-$ only. That of $Ti_5H_m^-$ is a sum of contributions from $Ti_5H_8^-$, $Ti_5H_9^-$ and $Ti_5H_{10}^-$. The corresponding calculated intensity distributions are shown in Fig. 3.13b. A pattern similar to that of $Ti_2H_m^-$ indicates an uptake of intact H_2 molecules only and is observed

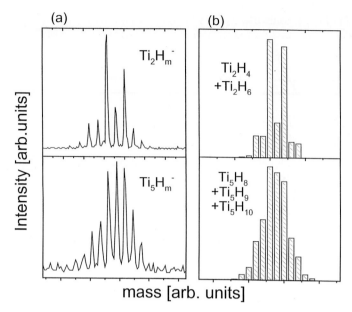

Fig. 3.13. (a) Two features observed in the mass spectrum of reacted $Ti_nH_m^-$ clusters. The two groups are assigned to $Ti_2H_m^-$ ($m = 4, 6$) and $Ti_5H_m^-$ ($m = 8, 9, 10$). (b) Simulation of the experimental data shown in (a). For agreement, for $Ti_2H_m^-$ only even m contribute, while for $Ti_5H_m^-$ a contribution from $Ti_5H_9^-$ with about 30% relative intensity is needed for fitting the experimental data (reproduced from [81]

for all smaller clusters with $n \leq 4$. All larger clusters show no preference for chemisorption of even m, as would be expected for dissociative chemisorption.

Fig. 3.14 displays photoelectron spectra of reacted $Ti_nH_m^-$ clusters with $n = 2-6$ [81]. The spectra of clusters with up to 4 Ti atoms exhibit two peaks (A,B) separated by about 1 eV. For the saturated species ($Ti_2H_6^-$, $Ti_3H_8^-$, $Ti_4H_8^-$), peak A is diminished. For larger $Ti_nH_m^-$ clusters, with $n > 4$, only a single peak C is visible. We also recorded spectra of some selected larger clusters (not shown) and found a similar pattern only with a single peak and no indication of a change of the electronic structure due to saturation: i.e., the pattern observed for $Ti_5H_m^-$ and $Ti_6H_m^-$ extends to larger cluster sizes.

Both, the mass and the photoelectron spectra reveal an structural transition of the $Ti_nH_m^-$ clusters occurring at $n > 4$. For the larger clusters with $n > 4$ the electronic structure has a certain similarity to the one of bulk TiH_2 and the small clusters seem to be different. Although these experimental results are no proof for a size dependent change from molecularly bound to dissociatively bound H_2 chemisorption, the data can be taken as a hint that such a transition may occur. Further studies are necessary and also a quantitative theoretical description is needed. However, such a transition

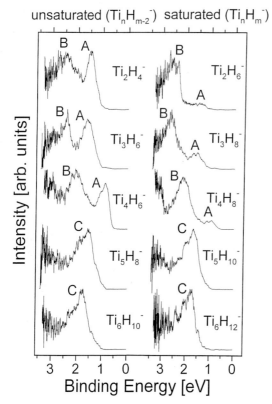

Fig. 3.14. Photoelectron spectra of $Ti_nH_m^-$ clusters recorded with a photon energy of 3.49 eV. The spectra of small clusters with $n \leq 4$ exhibit two peaks (A,B) with a strong decrease of intensity of peak A for the saturated species ($Ti_2H_6^-$, $Ti_3H_8^-$, $Ti_4H_8^-$). For larger clusters only a single feature (C) is observed (reproduced from [81])

might take place in many different chemisorption systems. A molecule, which is bound "almost" dissociatively to the cluster, has a weak internal bond and is highly "activated" for any chemical reaction. This "activation" of the molecule can explain the catalytic activity of the cluster independent of the detailed size and shape.

3.4.4 Example: Dynamics

The combination of the PES$^-$ technique with a femtosecond laser system has made it possible to study fast dynamical processes in mass-selected clusters [37–41]. The pioneering experiment has been performed in D. Neumark's group on the photodissociation of I_2^- [37,38]. The ultrashort laser pulse with a typical length of 100 fs is divided into two pulses: the pump and the probe pulse and the two pulses are delayed relatively to each other. First, the absorption of a photon from the pump pulse triggers a fast process like photodissociation or photoexcitation. As a result, the cluster anion dissociates or relaxes through other channels like autodetachment, internal vibrational redistribution or emission of photons. With the probe pulse a photoelectron spectrum at a selected delay is recorded and a series of spectra with increasing delay reveals the time evolution of the system. For example, in case of photodissociation at zero delay a spectrum of the intact cluster anion and at large delay the spectrum of the fragment anions is recorded (only the negatively-charged fragments are "visible" because of the high ionization potentials of any neutral fragments). New information is gained at delays in between, which reveals how the process takes place and whether new intermediate electronic states are populated.

As an example, the results of a time resolved PES$^-$ experiment on the photodissociation of Au$_3^-$ is discussed in the following [39]. For the time-resolved experiment the single UV laser pulse (Fig. 3.5) is replaced by two femtosecond pulses with a photon energy of 3.0 eV. The pulse widths of the two pulses are 160 fs corresponding to a time-resolution of about 230 fs.

In the case of Au$_3^-$ the competing channel of direct photodetachment is not allowed because of the high electron affinity of Au$_3^-$ (3.9 eV [23]). The dominant process after absorption of a 3 eV photon is photodissociation and there are two pathways [23]:

$$Au_3^- + h\nu \longrightarrow Au_1^- + Au_2 \qquad (3.1)$$
$$Au_3^- + h\nu \longrightarrow Au_2^- + Au_1 \qquad (3.2)$$

The additional electron can stay either with the atom or with the dimer fragment. For comparison, Fig. 3.15 displays photoelectron spectra of all species (Au$_1^-$, Au$_2^-$, Au$_3^-$), which possibly contribute to the probe photoemission signal [22]. These have been obtained using a standard ArF excimer laser at

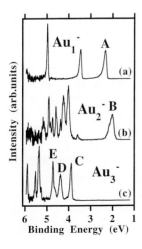

Fig. 3.15. Photoelectron spectra of Au$_1^-$ (a), Au$_2^-$ (b) and Au$_3^-$ (c) obtained using an ArF excimer laser with a photon energy of $h\nu = 6.4$ eV. For a discussion of the marked features see text (reproduced from [22])

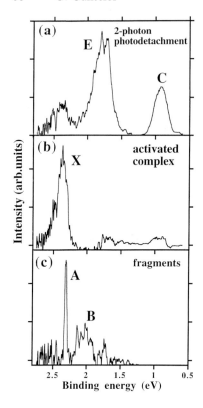

Fig. 3.16. Pump-probe photoelectron spectra of Au_3^- recorded with a photon energy of 3 eV (both pulses). Trace (a) shows a spectrum recorded with zero delay between the pulse corresponding to a single pulse with higher intensity. Trace (c) shows the signal recorded at long delays (> 2 ns). Trace (b) displays the spectrum of the activated complex Au_3^{-*}, which has a lifetime of 1.5 ns (from [39])

a photon energy of $h\nu = 6.4$ eV. The spectra give an overview of the electronic structure of the electronic ground states of Au_3^- (EA = 3.9 eV), Au_2^- (EA = 2.0 eV) and Au_1^- (EA = 2.3 eV).

Figure 3.16a displays a pump-probe photoelectron spectrum of Au_3^- recorded at zero delay. This spectrum can be assigned to a 2-photon photodetachment process: the first photon excites the Au_3^- into an electronically excited state and the second photon detaches an electron from the excited Au_3^-. Since there is no time for a geometrical rearrangement of the nucleii, the final states, which can be accessed by this process, are the same as in the case of a 1-photon process with the double photon energy. Therefore, the two peaks (E,C) observed in Fig. 3.16a can be assigned to transitions into the same final states of neutral Au_3 as in Fig. 3.15c. The BEs of features E and C in Fig. 3.16a correspond to the BEs of the peaks E and C in Fig. 3.15a, if 3 eV BE for the second photon are added. Feature D (Fig. 3.15c) is not observed in the 2-photon spectrum (Fig. 3.16a), which can be explained by the different selections rules valid for 2-photon photodetachment.

"Zero delay" corresponds to a single pulse and, therefore, the spectrum shown in Fig. 3.16a is always superimposed on any pump-probe spectrum recorded at larger delay; i.e., from the pump-probe spectra recorded at non-zero delay this "background" zero-delay spectrum has to be subtracted.

At large delays between the two pulses the spectrum of the two fragment anions Au_1^- and Au_2^- appear (Fig. 3.16c). Because of the small photon energy of $h\nu = 3\,\text{eV}$, only one peak is visible for each fragment (see Fig. 3.15). For the monomer, a sharp line at 2.3 eV binding energy (= electron affinity) (feature A) and for the dimer a broad feature centered at 2 eV binding energy (B) appear.

In the time regime up to about 1ns delay a new feature is observed (Fig. 3.16b). The new peak at about 2.4 eV BE is assigned to a long-lived excited state of Au_3^- (noted Au_3^{-*}), which is temporarily populated during the fragmentation process. Fig. 3.16b shows a photoelectron spectrum of this state, which is an "activated complex". It can only be observed using time-resolved PES$^-$ and its observation demonstrates the power of the new method.

According to the data above, the photofragmentation process occurs via an excited electronic state of Au_3^- with a lifetime of about 1.5 ns [39]. The state has a geometry different from the one of the anion ground state and decays by fragmentation. Thus, equations (1) and (2) have to be replaced by:

$$Au_3^- + h\nu \longrightarrow Au_3^{-*} \longrightarrow Au_1^- + Au_2$$
$$\longrightarrow Au_2^- + Au_1 \qquad (3.3)$$

Compared to the spectra recorded at 6.4 eV (Fig. 3.15) the pump-probe photoelectron spectra recorded at 3 eV (Fig. 3.16) contain less information. An improvement would be a considerably higher photon energy of the probe pulse and a tunable photon energy of the pump pulse. The latter should be tunable, because the processes, which are triggered by photoexcitation, depend on the photon energy. With such a setup, many different dynamical processes in clusters can be studied like the electron-hole recombination in semiconducting clusters or the dynamics of chemisorption of, e.g., CO adsorbed on transition metal clusters.

3.5 Conclusion and Outlook

Photoelectron spectroscopy of mass-selected anions, which has started about 15 years ago as a highly specialized technique, nowadays became a common and powerful tool for the spectroscopy of molecules, radicals, clusters, nanostructures and particles. In the very beginning this technique was mostly used for determination of the electronic ground states of elemental and molecular clusters and only the uppermost occupied valence orbitals could be observed. More recently, there were many improvements concerning the photon energy, the energy resolution and the mass resolution. With these improvements, even vibrational spectroscopy giving additional information about the geometric structure can be obtained. Also, various reacted clusters can be studied and the chemical properties of the small particles will be analysed. In the future, the combination with ultra-short laser pulses will shed light on the dynamical properties of clusters, which are almost unknown yet. The results of these studies of free clusters in the gas phase will be used as a guideline for the development of new catalysts and new cluster materials or new possible applications of the surprising properties of mass-selected clusters and nanostructures.

Acknowledgements. E. Recknagel and W. Eberhardt are gratefully acknowledged for their fruitful and permanent support.

References

1. M.S. Dresselhaus, G. Dresselhaus and P.C. Eklund, "Science of Fullerenes and Carbon Nanotubes", Academic Press, New York, 1995.
2. H.W. Kroto, Science **242**, 1139 (1988).
3. W.A. de Heer, W.D. Knight, M.Y. Chou, M.L. Cohen, Phys. **40**, 94 (1987).
4. W. Ekart and Z. Pensar, Phys. Rev. B **38**, 4273 (1988).
5. J. Berkowitz, "Photoabsorption, Photoionization and Photoelectron Spectroscopy", Academic Press, New York (1979).
6. G. Ertl and J. Küppers, "Low Energy Electron and Surface Chemistry", VCH-Verlag, Weinheim (1985).
7. D.G. Leopold, J.H. Ho and W.C. Lineberger, J. Chem. Phys. **86**, 1715 (1987).
8. K.M. Ervin and W.C. Lineberger, "Photoelectron Spectroscopy of Molecular Anions", in "Advances in Gas Phase Ion Chemistry", Vol. **1**, pp. 121–166 (1992)
9. K.M. McHugh, J.G. Eaton, G.H. Lee, H.W. Sarkas, L.H. Kidder, J.T. Snodgrass, M.R. Manaa and K.H. Bowen, J. Chem. Phys. **91**, 3792 (1989).
10. O. Cheshnovsky, S.H. Yang, P.L. Pettiette, M.J. Craycraft and R.E. Smalley, Rev. Sci. Instrum. **58**, 2131 (1987).
11. G. Ganteför, K.H. Meiwes-Broer and H.O. Lutz, Phys. Rev. A **37**, 2716 (1988).
12. D.W. Arnold, S.E. Bradforth, T.N. Kitsopoulos and D.M. Neumark, J. Chem. Phys. **95**, 8753 (1991).
13. M.F. Jarrold and J.E. Bower, J. Chem. Phys. **98**, 2399 (1993).
14. G. v. Helden, N.G. Gotts and M.T. Bowers, Nature **363**, 60 (1993).
15. Kai-Ming Ho, A.A. Shvartsburg, B. Pan, Zhong-Yi Lu, Cai-Zhuang Wang, J.G. Wacker, J.L. Fye and M.F. Jarrold, Nature **392**, 582 (1998).
16. G. von Helden, M.T. Hsu, N. Gotts and M.T. Bowers, J. Phys. Chem. **97**, 8182 (1993).
17. S. Yang, C.L. Pettiette, J. Conceicao, O. Cheshnovsky and R.E. Smalley, Chem. Phys. Lett. **139**, 233 (1987).
18. H. Handschuh, G. Ganteför, B. Kessler, P.S. Bechthold and W. Eberhardt, Phys. Rev. Lett. **74**, 1095 (1995).
19. O. Gunnarsson, H. Handschuh, P.S. Bechthold, B. Kessler, G. Ganteför and W. Eberhardt, Phys. Rev. Lett. **74**, 1875 (1995).
20. H. Kietzmann, R. Rochow, G. Ganteför, W. Eberhardt, K. Vietze, G. Seifert, P.W. Fowler, Phys. Rev. Lett. **81**, 5378 (1998).
21. H. Handschuh, Chia-Yen Cha, P.S. Bechthold, G. Ganteför and W. Eberhardt, J. Chem. Phys. **102**, 6406 (1995).

22. H. Handschuh, G. Ganteför, P.S. Bechthold and W. Eberhardt, J. Chem. Phys. **100**, 7093 (1994).
23. J. Ho, K.M. Ervin and W.C. Lineberger, J. Chem. Phys. **93**, 6987 (1990).
24. H. Handschuh, G. Ganteför and W. Eberhardt, Rev. Sci. Instrum. **66**, 3838 (1995).
25. C.E. Moore, "Atomic Energy Levels", Natl. Bur. Stand., Ref. Data Ser. 35 (Natl. Bur. Stand., Washington D.C. 1971).
26. H. Hotop and W.C. Lineberger, J. Phys. Chem. Ref. Data **14**, 731 (1985).
27. P. Xia and L.A. Bloomfield, Phys. Rev. Lett. **70**, 1779 (1993).
28. M.J. DeLuca, Baohua Niu and M. A Johnson, J. Chem. Phys. **88**, 5857 (1988).
29. R. Busani, M. Folkers and O. Cheshnovsky, Phys. Rev. Lett. **81**, 3836 (1998).
30. H. Yoshida, A. Terasaki, K. Kobayashi, M. Tsukada and T. Kondow, J. Chem. Phys. **102**, 5960 (1995).
31. M. Gausa, R. Kaschner, G. Seifert, J.H. Faehrmann, H.O. Lutz and K.-H. Meiwes-Broer, J. Chem. Phys. **104**, 9719 (1996).
32. A. Nakajima, T. Taguwa, K. Nakao, M. Gomei, R. Kishi, S. Iwata and K. Kaya, J. Chem. Phys. **103**, 2050 (1995).
33. M. Kohno, S. Suzuki, H. Shiromanu, T. Moriwaki and Y. Achiba, Chem. Phys. Lett. **282**, 330 (1998).
34. Lai-Cheng Wang and Hansong Cheng, Phys. Rev. Lett. **78**, 2983 (1997).
35. Xue-Bin Wang, Chuan-Fan Ding and Lai-Sheng Wang, Phys. Rev. Lett. **81**, 3351 (1998).
36. P. Kruit and F.H. Read, J. Phys. **E16**, 313 (1983).
37. B.J. Greenblatt, M.T. Zanni and D.M. Neumark, Chem. Phys. Lett. **258**, 523 (1996).
38. B.J. Greenblatt, M.T. Zanni and D.M. Neumark, Science **276**, 1675 (1997).
39. G. Ganteför, S. Kraus and W. Eberhardt, J. Electr. Spectr. Rel. Phen. **35**, 88–91, (1997).
40. N. Pontius, P.S. Bechthold, M. Neeb and W. Eberhardt, Phys. Rev. Lett. **84**, 1132 (2000).
41. S. Minemoto, J. Müller, G. Ganteför, H.J. Münzer, J. Boneberg and P. Leiderer, Phys. Rev. Lett. **84**, 3554 (2000).
42. S. Burkart, N. Blessing, B. Klipp, J. Müller, G. Ganteför and G. Seifert, Chem. Phys. Lett. **301**, 546 (1999).
43. G. Ganteför, H.R. Siekmann, H.O. Lutz and K.H. Meiwes-Broer, Chem. Phys. Lett. **165**, 293 (1990).
44. H.R. Siekmann, Ch. Lüders, J. Fährmann, H.O. Lutz, K.H. Meiwes-Broer, Z. Phys. D **20**, 417 (1991).
45. Chia-Yen Cha, G. Ganteför and W. Eberhardt, Rev. Sci. Instrum. **63**, 5661 (1992).
46. Bu. Wrenger and K.H. Meiwes-Broer, Rev. Sci. Instrum. **68**, 2027 (1997).
47. V.E. Bondebey and J.H. English, J. Chem. Phys. **74**, 6978 (1981).
48. T.G. Dietz, M.A. Duncan, D.E. Powers and R.E. Smalley, J. Chem. Phys. **74**, 6511 (1981).
49. B.A. Mamyrin, V.I. Karataev, D.V. Shmikk and V.A. Zagulin, Sov. Phys. JETP **37**, 45 (1973).
50. W.C. Wiley and I.H. Mclaren, Rev. Sci. Instrum. **26**, 1150 (1955).
51. T. Bergmann, T.P. Martin and H. Schaber, Rev. Sci. Instrum. **60**, 792 (1989).
52. R.B. Metz, A. Weaver, S.E. Bradforth, T.N. Kitsopoulos and D.M. Neumark, J. Phys. Chem. **94**, 1377 (1990).

53. Cangshan Xu, G.R. Burton, T.R. Taylor and D.M. Neumark, J. Chem. Phys. **107**, 3428 (1997).
54. W. de Heer, Rev. Mod. Phys. **65**, 611 (1993).
55. W.D. Knight, K. Clemenger, W.A. de Heer, W.A. Saunders, M.Y. Chou and M.L. Cohen, Phys. Rev. Lett. **52**, 2141 (1984).
56. O. Cheshnovsky, K.J. Taylor, J. Conceicao and R.E. Smalley, Phys. Rev. Lett. **64**, 1785 (1990).
57. K.J. Taylor, C.L. Pettiette-Hall, O. Cheshnovsky and R.E. Smalley, J. Chem. Phys. **96**, 3319 (1992).
58. G. Ganteför, M. Gausa, K.H. Meiwes-Broer and H.O. Lutz, Faraday Discuss. Chem. Soc. **86**, 1 (1988).
59. Chia-Yen Cha, G. Ganteför and W. Eberhardt, J. Chem. Phys. **99**, 6308 (1993).
60. Chia-Yen Cha, G. Ganteför and W. Eberhardt, Z. Phys. D **26**, 307 (1993).
61. H. Handschuh, Chia-Yen Cha, H. Möller, P.S. Bechthold, G. Ganteför and W. Eberhardt, Chem. Phys. Lett. **227**, 496 (1994).
62. V. Bonacic-Koutecky, L. Cespiva, P. Fantucci, J. Pittner and J. Koutecky, J. Chem. Phys. **100**, 490 (1994).
63. J. Müller, Bei Liu, A.A. Shvartsburg, S. Ogut, J.R. Chelikowsky, K.W.M. Siu, Kai-Ming Ho and G. Ganteför, Phys. Rev. Lett. **85**, 1666 (2000).
64. K. Raghavachari and C.M. Rohlfing, J. Chem. Phys. **94**, 3670 (1991).
65. M.D. Morse, M.E. Geusic, J.R. Heath and R.E. Smalley, J. Chem. Phys. **83**, 2293 (1985).
66. M.R. Zakin, R.O. Brickman, D.M. Cox and A. Kaldor, J. Chem. Phys. **88**, 3555 (1988).
67. J.L. Elkind, F.D. Weiss, J.M. Alford, R.T. Laaksonen and R.E. Smalley, J. Chem. Phys. **88**, 5215 (1988).
68. Yoon Mi Hamrick and M.D. Morse, J. Phys. Chem. **93**, 6494 (1989).
69. M.R. Zakin, R.O. Brickman, D.M. Cox and A. Kaldor, J. Chem. Phys. **88**, 6605 (1988).
70. C. Berg, T. Schindler, G. Niedern-Schatteburg and V.E. Bondybey, J. Chem. Phys. **102**, 4870 (1995).
71. R.L. Whetten, D.M. Cox, D.J. Trevor and A. Kaldor, Phys. Rev. Lett. **54**, 1494 (1985).
72. H. Kietzmann, J. Morenzin, P.S. Bechthold, G. Ganteför and W. Eberhardt, J. Chem. Phys. **109**, 2275 (1998).
73. U. Heiz, A. Sanchez, S. Abbet and W.-D. Schneider, J. Am. Chem. Soc. **121**, 3214 (1999).
74. A. Sanchez, A. Abbet, U. Heiz, W.-D. Schneider, H. Häkkinen, R.N. Barnett and U. Landman, J. Phys. Chem. A **103**, 9573 (1999).
75. K.-H. Meiwes-Broer, Hyperfine Interact. **89**, 263 (1994).
76. R. Hoffman, Rev. Mod. Phys. **60**, 601 (1988).
77. P. Cremaschi and J.L. Whitten, Phys. Rev. Lett. **46**, 1242 (1981).
78. "Organometallic Ion Chemistry", Ben Freiser (Ed.), Kluever Academic Publishers, Dordrecht (1996).
79. K.P. Huber and G. Herzberg, "Molecular Spectra and Molecular Structure: IV. Constants of Diatomic Molecules", Van Nostrand Reinhold Company, New York, 1979.
80. P. Nordlander, S. Holloway and J.K. Norskov, Surf. Sci. **136**, 59 (1984).
81. S. Burkart, N. Blessing and G. Ganteför, Phys. Rev. B **60**, 15639 (1999).

4 Quantum Tunneling of the Magnetization in Molecular Nanoclusters

R. Sessoli, D. Gatteschi and W. Wernsdorfer

4.1 Introduction

The quantum tunneling of the magnetization is a typical mesoscopic effect located at the transition from classical to quantum physics. Molecular clusters are small objects characterized by a small spin compared to single domain particles but significantly larger than the largest spin observable in atoms [1]. In this view it is not surprising that the magnetic properties of molecular clusters have a marked quantum character. On the other side the macroscopic magnetization of these clusters shows bi-stability, hysteresis effects and superparamagnetic behavior [2] as the more classical single-domain particles, so that they are also known as "single molecule magnets". Molecular clusters are thus placed at the interface between the classic and the quantum domains and have provided an unexpected richness of quantum effects, which have attracted continuously increasing interest [3]. They have the outstanding advantage of being well-characterized systems with identical objects well ordered in a crystal lattice. Physicists are often interested only in the magnetic core of the cluster, relegating the organic part to the role of a diamagnetic matrix. The solution coordination chemistry technique, used to obtain these clusters, is an important feature, as it allows a fine tuning of the structure and physical properties of the clusters [4–7]. Most of the physical studies have been concentrated on two system, a dodecanuclear manganese [8] and an octanuclear iron cluster [9] both characterized by spin $S = 10$, but we are currently observing an exponentially increasing number of magnetic clusters that are synthesized. The record spin is now of $S = 51/2$ for a cluster containing nine Mn(II) ions, $S = 5/2$, and six Mo(V), $S = 1/2$, ferromagnetically coupled spins [10]. We can reasonably expect that the temperature at which the magnetic hysteresis is observed increases as clusters with higher spin are obtained. Molecular magnetic clusters will therefore also attract interest for potential applications as single molecule memory units. For instance one of the most fascinating quantum effects observed in these systems, the topological interference or Berry phase in the tunneling pathways, is an example of transverse field assisted writing on a magnetic memory, a very hot topic in materials science.

The aim of this contribution to the book is that of providing a non-specialist reader with an overview of the most important experimental evidences of the quantum effects in the dynamics of the magnetization of molecular clusters even if space limitation does not allow a comprehensive treatment of the subject. The two systems reported here are $[Mn_{12}O_{12}(CH_3COOH)_{16}(H_2O)_4]$, $Mn_{12}ac$, and $[Fe_8O_2(OH)_{12}(tacn)_6]Br_8 9H_2O$, Fe_8Br, whose structures are schematized in Fig. 4.1. They are characterized by a large spin ground state $S = 10$ arising from antiferromagnetic interaction, mediated mainly by the bridging oxygen and hydroxy groups. The spin topology of the eight $S = 5/2$ spins in Fe_8Br gives rise to non-compensation and to the ferrimagnetic spin structure schematized by the arrows [11] in Fig. 4.1, while for $Mn_{12}ac$, non compensation is also due to the different value of the spin of the Mn(IV) ions, $S = 3/2$ and that of the Mn(III) ions, $S = 2$ [12]. The charge and the spin are in both cases well localized on the metal centers. From the chemical point of view the main difference between the two systems is that $Mn_{12}ac$ is a neutral molecule, stable in solution, which can be chemically modified. A large series of derivatives also has been obtained with a different number of unpaired electrons [13–15]. On the contrary, there are no evidences that the octanuclear iron cluster exists in solution and only minor structural changes, like the partial substitution of the bromide counterions, have been possible [16]. From a physical point of view the main difference between the two systems is the symmetry, which is strictly tetragonal in $Mn_{12}ac$, while Fe_8Br has no symmetry at all crystallizing in the triclinic P1 space group.

In order to observe the slow relaxation of the magnetization of molecular clusters, a sizeable barrier for the reorientation of the magnetization must be present and large ground spin state and an easy axis type magnetic anisotropy

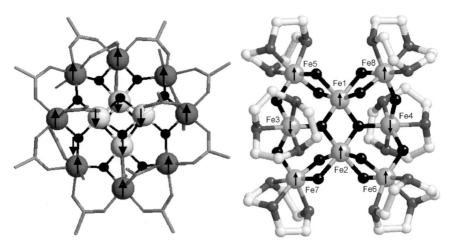

Fig. 4.1. View of the structure of $Mn_{12}ax$ (*left*) and Fe_8Br (*right*). The metal ions are represented by large spheres, which in $Mn_{12}ac$ are darker for the Mn(III) ions. The bridging oxygen atoms are drawn as small black spheres. The arrows represent the spin structure as suggested by calculation and confirmed by polarized neutron diffraction experiments

4 Quantum Tunneling of the Magnetization in Molecular Nanoclusters 57

are therefore necessary, as observed in both Mn_{12}ac and Fe_8Br. In order to observe quantum phenomena it is necessary that the $\pm M$ levels, eigenstates of S_z, are suitably split (tunnel splitting) by a transverse field, which can be either intrinsic (associated with the transverse anisotropy of the cluster) or extrinsic. The magnetic anisotropy therefore plays a fundamental role in determining the properties of the clusters. A discussion on the origin of the magnetic anisotropy will be given in the next section. In Sect. 4.3 we will briefly describe the slow relaxation of the magnetization due to spin-phonon coupling. Quantum effects as the resonant tunneling are the object of Sect. 4.4 while Sect. 4.5 is devoted to the observation of the topological quenching of the tunneling for the application of a transverse field. The role played by the dipolar field of the neighboring clusters is the subject of Sect. 4.6. The recently observed isotopic effect due to the hyperfine field is a further confirmation of the tunneling picture and will be presented in Sect. 4.7.

4.2 The Magnetic Anisotropy of Molecular Clusters

The magnetic anisotropy of molecular clusters is determined by two sets of contributions, one originating from the individual building blocks of the cluster (single ion anisotropy), and the other from the interaction between the building blocks (spin-spin anisotropy)[4,17]. A convenient way of expressing these terms is by using a spin hamiltonian, i.e. a parametric approach in which all the variables are substituted by spin variables, and the physically relevant parameters are obtained from the analysis of experimental data. In this scheme the single ion contribution can be written as:

$$H_{si} = \sum_{i,n,k} B_{i,n}^k O_{i,n}^k ,\qquad(4.1)$$

where the sum is extended to all the individual spins $S_i > 1/2$. The $O_{i,n}^k$ operators correspond to S_i^n power angular momentum operators, with n even, and are listed in Table 4.1 up to the 6th power [18]. The maximum value of n to be included in (4.1) is $2S_i$, k values to be included range from $-n$ to n, the actual values depending on the site symmetry. $B_{i,n}^k$ are suitable parameters. Their physical origin is essentially crystal field effects mediated by spin orbit coupling for transition metal and rare earth ions, while for organic radicals the dipolar interaction between the unpaired electrons is dominant. The second order terms, $n = 2$, are usually the dominant ones. They are often written in the simpler formalism:

$$H = \sum_i D_i[\boldsymbol{S_{i,z}}^2 - S(S+1)/3] + E_i(\boldsymbol{S_x}^2 - \boldsymbol{S_y}^2) .\qquad(4.2)$$

The spin-spin contributions to the magnetic anisotropy depending on the interaction between magnetic centers can be of both through space (magnetic

Table 4.1. Equivalent operators for the polynomials expression of the magnetic anisotropy in (4.1). $\{A,B\}^\otimes$ is used as a shorthand for $1/2(AB+BA)$

$n=2$	$O_2^0 = 3\mathbf{S}_z^2 - S(S+1)$
	$O_2^2 = 1/2(\mathbf{S}_+^2 + \mathbf{S}_-^2)$
$n=4$	$O_4^0 = 35\mathbf{S}_z^4 - 30S(S+1)\mathbf{S}_z^2 + 25\mathbf{S}_z^2 - 6S(S+1) + 3S^2(S+1)^2$
	$O_4^2 = 1/2\,\{\,(7\mathbf{S}_z^2 - S(S+1) - 5), (\mathbf{S}_+^2 + \mathbf{S}_-^2)\,\}^\otimes$
	$O_4^3 = 1/2\,\{\,\mathbf{S}_z, (\mathbf{S}_+^3 + \mathbf{S}_-^3)\,\}^\otimes$
	$O_4^4 = 1/2\,(\mathbf{S}_+^4 + \mathbf{S}_-^4)$
$n=6$	$O_6^0 = 231\mathbf{S}_z^6 - 315S(S+1)\mathbf{S}_z^4 - 735\mathbf{S}_z^4 + 105S^2(S+1)^2\mathbf{S}_z^2$
	$\qquad - 525S(S+1)\mathbf{S}_z^2 + 294\mathbf{S}_z^2 - 5S^3(S+1)^3 + 40S^2(S+1)^2 - 60S(S+1)$
	$O_6^3 = 1/2\,\{\,(11\mathbf{S}_z^3 - 3S(S+1)\mathbf{S}_z - 59\mathbf{S}_z), (\mathbf{S}_+^3 + \mathbf{S}_-^3)\,\}^\otimes$
	$O_4^3 = 1/2\,\{\,(11\mathbf{S}_z^2 - S(S+1) - 38), (\mathbf{S}_+^4 + \mathbf{S}_-^4)\,\}^\otimes$
	$O_4^4 = 1/2\,\{\,(\mathbf{S}_+^6 + \mathbf{S}_-^6)\,\}$

multipolar) and through bond (via spin orbit coupling) nature. The effects to which these terms give rise are similar to each other; therefore it is in general difficult to effectively recognize from the observed anisotropy which is the physical origin of the interaction. Developing the terms up to the second order, the hamiltonian appropriate to describe these interactions can be written as:

$$H = \boldsymbol{S}_i \cdot \boldsymbol{D}_{ij} \cdot \boldsymbol{S}_j + \boldsymbol{d}_{ij} \cdot (\boldsymbol{S}_i \times \boldsymbol{S}_j)\,, \tag{4.3}$$

where the second term is the so-called Dzyaloshinsky-Moriya term [19], which vanishes for centrosymmetric pairs. $\boldsymbol{D}_{ij}$ is a symmetric traceless tensor, while $\boldsymbol{d}_{ij}$ is a vector. The isotropic $J_{ij}\boldsymbol{S}_i \cdot \boldsymbol{S}_j$ term has not been included in 4.3 because it does not affect the anisotropy. Higher order terms must in principle be taken into consideration and, especially when the individual S_i spins are large, they may give important contributions.

The calculation of the magnetic anisotropy for large clusters using the hamiltonians (4.1–4.3) is impossible, because of the huge size of the matrices to be solved. In fact for a cluster of N individual spins S_i the matrix is $(2S_i+1)^N$, which becomes rapidly totally intractable. The size of the matrix can be reduced if only isotropic terms of the interaction are used. In this case the total spin S is a good quantum number and the size of the matrices can be suitably reduced.

In practice quite often a simplified treatment is used, in which only the ground total spin state multiplet, S, is considered, and the effect of the anisotropic operators is considered only on the basis of $2S+1$ states corresponding to this spin value. In this case the appropriate hamiltonian can be written as:

$$H_S = \sum_{n,k} B_n^k O_n^k , \tag{4.4}$$

where the symbols have the same meaning as in (4.1). The allowed values of n are no longer restricted only to be even, if the effects of antisymmetric terms, like the Dzyaloshinsky-Moriya terms must be taken into consideration. The values of the B_n^k parameters in the coupled representation can be related to the single ion and spin-spin interaction parameters provided that isotropic exchange terms can be assumed to be dominant [20].

Detailed attempts to correlate the individual anisotropic terms with the anisotropy observed in the total spin states, have been made in the case of relatively simple clusters, like the hexanuclear rings comprising iron(III) ions [4,21]. The main result has been that of showing that dipolar interactions alone are enough to justify the observed anisotropy in some cases, while in others single ion contributions play a dominant role.

For $Mn_{12}ac$ and Fe_8Br, several attempts have been made to obtain reliable values of the anisotropic spin hamiltonian parameters using several different techniques (see [22] and [23] for a review on the experimental techniques for the determination of magnetic anisotropy in molecular clusters). The energy splitting of an $S = 10$ multiplet due to the axial anisotropy of second order is shown in Fig. 4.2. A double well formalism is used in which M levels of different sign are localized in different potential curves. $Mn_{12}ac$ has a crystal imposed S_4 symmetry, which requires the introduction of the O_n^0 and O_n^4 terms in the spin hamiltonian (4.4), if the antisymmetric terms are neglected. The O_n^0 operators give only diagonal matrix elements, thus leaving the $\pm M$, M being projection of the spin along z, levels degenerate in pairs. The degeneracy can be lifted by the O_n^4 operator that mixes states differing

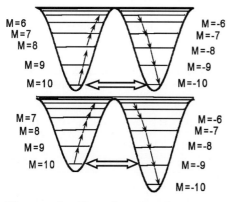

Fig. 4.2. Double well potential due to the axial magnetic anisotropy experienced by the $S = 10$ spin clusters, in zero longitudinal field (*top*) and in a longitudinal field that brings the M and $(-M + 1)$ states in energy coincidences

in M by ± 4. Therefore, in the ground $S = 10$ state, it provides an effective transverse field splitting of the $M = \pm 2$ degenerate doublet in first order. However it can split the lowest $M = \pm 10$ doublet only in fifth order, thus providing only an exceedingly small tunnel splitting. Further it can split only the $M = \pm k$, with k even, doublets. An important aspect of the S_4 symmetry of Mn$_{12}$ac is that it lacks an inversion center, so that antisymmetric exchange term may in principle be included. These allow the presence of odd operators, which may produce additional splitting of the S multiplets. This possibility has so far not been extensively explored, but recently it has been proposed to justify the lack of parity effect in the hysterisis loop of Mn$_{12}$ac [24].

The inversion symmetry is lacking also in Fe$_8$Br, which in fact has no symmetry at all. However it has been customary to assume a higher symmetry, namely D_2, for the calculation of the spin levels in the limit of isotropic exchange. This same symmetry has been used for the inclusion of anisotropy in the ground $S = 10$ state. In this approximation the even O_n^0, O_n^2, and O_n^4 operators have been taken into consideration. The coefficient of the O_2^2 operator is quite large, see Table 4.2, and this provides a sizeable transverse field which determines a large splitting and a large admixture of the $\pm M$ levels with $|M| < 6$, in such a way that the double well formalism as shown in Fig. 4.2 loses significance [25,26]. The first attempts to describe the zero field splitting through the analysis of the polycrystalline powder EPR spectra of Mn$_{12}$ac [27] and Fe$_8$Br [28], were limited to the minimal number of parameters, including only the second order parameters D and E (the latter is zero by symmetry for Mn$_{12}$ac. More detailed EPR spectra [29], taking advantage of single crystal data, showed that fourth order terms had to be included to explain the details [25]. In particular it was apparent that the separation between the resonance fields of neighboring $M \longrightarrow M + 1$ transitions was not independent of M as it would be required by the hamiltonian (4.2). However the experimental determination of the fourth order parameters is by no means simple, because the parameters are highly correlated. A slightly different set of parameters has been given by different authors [30]. Recently it has also been possible to perform EPR experiments in zero field [31], by

Table 4.2. Spin-Hamiltonian experimental parameters, as defined in (4.1–4.4) for Mn$_{12}$ac and Fe$_8$Br. The values are in K. HFEPR stays for High Field EPR data and INS for Inelastic Neutron Scattering data

	D	E	B_4^0	B_4^2	B_4^4	$\Delta E_{\pm 10}$	Ref.
Mn$_{12}$ac	−0.66	0	-3.2×10^{-5}	0	$\pm 6 \times 10^{-5}$	1.1×10^{-11}	HFEPR [29]
	−0.658	0	-3.35×10^{-5}	0	$\pm 4.3 \times 10^{-5}$	2.6×10^{-12}	INS [35]
Fe$_8$Br	−0.295	0.055	2.3×10^{-6}	-7×10^{-6}	1.1×10^{-5}	6.5×10^{-9}	HFEPR [25]
	−0.292	0.046	1.0×10^{-6}	1.1×10^{-7}	9×10^{-6}	3.5×10^{-11}	INS [37]

scanning the microwave frequency, and torque magnetometry experiments have revealed to be a very precious tool for the determination of the axial anisotropy [32]. Inelastic neutron scattering is another important tool for obtaining the magnetic anisotropy of clusters. This technique has been applied to both Mn$_{12}$ac [33–35] and Fe$_8$Br [36,37], providing the parameters that are reported in Table 4.2 with those of HFEPR experiments.

The origin of the magnetic anisotropy of Mn$_{12}$ac is associated with the single ion anisotropy of the Jahn-Teller distorted manganese(III) ions [38], while for Fe$_8$Br it is a blend of single ion contributions and dipolar spin-spin interactions [25]. Recently an independent estimation of the axial magnetic anisotropy in Mn$_{12}$ac has been provided by electronic structure calculations using the density functional approach [39,40].

4.3 The Superparamagnetic Behavior

The first evidences of an anomalous slow relaxation of the magnetization of Mn$_{12}$ac came from ac magnetic susceptibility measurements [27]. It is well known that if an ac field is applied in zero static field a paramagnet gives only an in-phase response [41]. This is true if frequencies not much higher than the MHz are used, because in this case the spin-spin mechanism, which has a time scale of the order of 10^{-9} s in concentrated samples and is essentially temperature independent, is responsible for the relaxation. As soon as a field is applied the spin-spin relaxation is no longer efficient and the spin-lattice relaxation, which is very slow at low temperature, gives rise to an out of phase signal when the relaxation time τ is ca. ω^{-1}. The experiments performed in zero static field on Mn$_{12}$ac revealed an almost single-exponential decay [42], with the relaxation time following an Arrhenius law:

$$\tau = \tau_0 exp(-\Delta/k_\mathrm{B} T) , \qquad (4.5)$$

with $\tau_0 = 2 \times 10^{-7}$ s and $\Delta/k_\mathrm{B} = 61$ K. By decreasing the temperature below 3 K the relaxation becomes so slow that the decay of the remanent magnetization can be directly monitored and a hysteresis of dynamical origin is observed.

Mn$_{12}$ac was the first paramagnet showing out of phase ac susceptibility in the absence of an external static field. In fact the unusual high spin of the ground state of Mn$_{12}$ac, combined with the magnetic anisotropy that stabilizes the $M = \pm 10$ states, gives rise to the energy barrier for the reorientation of the magnetization.

The crossing of the barrier, which is the process that brings the magnetization to the equilibrium value after removing the field, can be described as a multi-step Orbach mechanism driven by spin-phonon interaction. In fact the two states are not connected through one excited state but through $(2S + 1) - 2$ states higher in energy. A complete treatment can be found in ref [43,44], and here we just summarize the main result. In the assumption of low temperature $T \ll \Delta/k_B$

$$1/\tau = \frac{3}{2\pi} \frac{|V_{10}|^2}{\hbar^4 \rho c^5} \left[\frac{\Delta}{S^2}\right] \frac{\exp[-\Delta/k_B T]}{1 - \exp[-\Delta/(k_B T S^2)]}, \tag{4.6}$$

where V_{10} is the matrix element of the spin-phonon interaction between the state $M = 0$ and $M = 1$ on the top of the barrier, which corresponds to the slowest part of the relaxation process, ρ is the density and c is the sound speed. It is interesting to notice that the pre-exponential factor is dependent on the spin value if we assume a constant value of Δ. Indeed the pre-exponential factor experimentally observed for $Mn_{12}ac$ and for Fe_8Br ($\tau_0 = 3.4 \times 10^{-8}$ s) is unusually long, compared to the value of 10^{-10}–10^{-11} s observed in classical superparamagnets [41]. In the case of $Mn_{12}ac$ the relaxation time is so long that the magnetization freezes at temperatures well above those of inter-cluster magnetic ordering. In fact for $Mn_{12}ac$ no ordering has been observed as it should be driven by inter-cluster dipolar interactions with a transition temperature estimated well below 1 K. Values of T_c equal to 0.17 K for Fe_8Br and 0.42 K for $Mn_{12}ac$ have been recently estimated [45]. At this temperature the relaxation time of the magnetization according to (4.5) is well above 10^{10}s and therefore too long to be measured.

Specific heat measurements, that have not revealed any anomaly in zero field [46], confirm that the slow relaxation of the magnetization is not due to magnetic ordering. Measurements have been performed by dissolving the clusters in organic solvents so that at low temperature the solvent forms a diamagnetic matrix [47,48]. These experiments showed that the slow relaxation, at least in the thermally activated regime, is essentially not-affected by inter-cluster interactions. The magnetic hysteresis measured through magnetic circular dichroism on such a glassy material has confirmed that the magnetic bistability is associated to the single cluster and is not a cooperative behavior. A datum can be therefore stored in a single-molecule magnetic memory unit [49].

The value of the experimentally determined barrier for $Mn_{12}ac$ is significantly smaller than the difference in energy between the $M = \pm 10$ and $M = 0$ calculated using the parameters in Table 4.1. We have already seen that the fourth order transverse magnetic anisotropy mixes the states characterized by $\Delta M = 4n$. Below-barrier transitions are therefore responsible of a short-cut of the barrier, giving rise to an Arrhenius behavior with a reduced height of the barrier [50]. This effect is even more evident in Fe_8Br where the states in the upper part of the barrier are strongly admixed.

4.4 Longitudinal Field Dependence of the Relaxation Rate: The Stepped Hysteresis

The underbarrier process, commonly called tunneling of the magnetization in agreement with a semiclassical picture of the barrier, is at the origin of one of the most striking quantum effects in the dynamics of the magnetization of molecular clusters: the reproducible steps observed in the hysteresis cycle [51,52], as shown for Fe_8Br in Fig. 4.3. As the hysteresis itself, the steps have a pure dynamic origin. The horizontal parts of the curve reported in Fig. 4.3 correspond to a condition of slow relaxation. This is the case when the application of an external longitudinal field removes the energy coincidence of the levels on the opposite sides of the barrier (see Fig. 4.2). In this case the transition probability of crossing the barrier through a tunneling mechanism between states that are thermally populated is strongly reduced and the time required for the reversal of the magnetization is very long. For some critical values of the external field, the energy levels on opposite sides of the double well potential are brought back to resonance, thus allowing a shortcut of the barrier. The relaxation time of the magnetization is strongly reduced and the system can more rapidly approach equilibrium, giving rise to a step in the magnetization. The amplitude of the steps strongly depends on the scan speed of the external field compared to the characteristic time. On the other hand the position of the steps is independent of the scan speed, at least in first approximation, being related only to the parameters of the spin hamiltonian mentioned before. If the spin hamiltonian describing the axial anisotropy contains only second order terms the steps occurs at $H_c = nD/g\mu_B$, where n is an integer. In this case the field at which the step occurs depends only on n

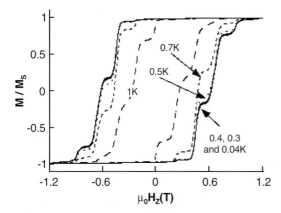

Fig. 4.3. Hysteresis loops measured on a single crystal of Fe_8Br by applying the magnetic field along the a crystallographic axis, which is at ca. $10°$ from the easy axis of magnetization, z. Steps are observed due to resonant tunneling. The curves becomes temperature independent below 0.36 K as a signature of pure tunneling between the lowest states in the well

as all the pairs characterized by $M = m$ and $M = -m + n$ are in resonance. For $Mn_{12}ac$ and Fe_8Br there are strong evidences that a second order spin hamiltonian is not adequate to describe the magnetic anisotropy. If the axial anisotropy contains also terms higher than the second order at a certain field only a pair of levels can be in resonance. If these additional terms complicate the picture a little, they have also permitted to identify the pairs of levels that are responsible for the tunneling at a certain temperature [53]. In fact as the temperature is decreased the thermal population of the upper levels is strongly reduced. Even if the tunneling between the lower levels in the potential well (those with larger $|M|$) has a smaller probability, its efficiency becomes stronger and stronger compared to the thermal activated mechanism as the temperature is reduced. It eventually leads to so called pure tunneling, which occurs within the ground pair $M = \pm S$. Chudnovsky et al. [54,55] have predicted that the transition from a thermally activated tunneling to pure tunneling is abrupt. They suggested that it occurs through a first order transition in zero transverse field and through a second order one in strong transverse field, and experiments have been performed to confirm this hypothesis. [53,56] The dynamic nature of the stepped hysteresis cycle also has been confirmed by other experiments not directly related to the magnetization. For instance anomalies in the specific heat [57] are observed at the critical fields as well as in the echo in ^{1}H NMR [58] experiments.

Up to now we have only presented the effects on the dynamics of the magnetization of the tunneling through the anisotropy barrier without mentioning the origin of this tunneling mechanism. Of course any term of the spin hamiltonian that does not commute with the axial anisotropy terms can admix the states on the opposite sites of the barrier giving rise to the tunneling. A transverse field of any origin, external, dipolar or nuclear, could in principle be responsible of the observed behavior. However a very simple calculation shows that, while its effect on the levels on the top of the barrier is relevant, the states with larger $|M|$ are practically unaffected. In a relatively small magnetic field (< 0.1 T) the extent of admixing of the $\pm S$ states, the so called tunnel splitting Δ, is really infinitesimal for both Fe_8Br and $Mn_{12}ac$, which have $S = 10$. The transverse magnetic anisotropy is indeed much more efficient in promoting tunneling between the lowest pairs of states. If only the magnetic anisotropy is taken into account, however, selection rules should be operative, depending on the symmetry of the systems. For instance for $Mn_{12}ac$, which has tetragonal symmetry, tunneling should be allowed only between states whose eigenvalues of $\boldsymbol{S}_z$ differ by $4n$, n being an integer. On the contrary for Fe_8Br this difference must be equal to $2n$ for tunneling to be active. Despite the fact that the experimental transverse magnetic anisotropy as shown above gives the right order of magnitude for the tunnel probability, violation of the selection rules is commonly observed. As we will show below other kinds of experiments have provided clearer evidence of the parity effect due to the symmetry of the magnetic anisotropy. In fact a small transverse field, which breaks the symmetry of the spin hamiltonian, also breaks the selection rules. The order of magnitude of the tunneling process is always provided by the magnetic anisotropy [50,59–61].

Tunneling can be evidenced at relatively high temperatures if it involves thermally populated states in the double well potential. For instance the ac susceptibility measurements have revealed oscillations in the real and imaginary components when a static field is applied. For Fe_8Br well defined oscillations are still observed at 7 K [62]. If the tunneling involves the ground states, it must be temperature independent. This has been observed below ca. 0.36 K in Fe_8Br as shown by the hysteresis loops or by the curves of decay of the magnetization [63], which become super-imposable below this temperature. On the contrary for $Mn_{12}ac$ the pure tunneling, at least in weak transverse field, is such a slow process that so far has not been experimentally determined.

4.5 Transverse Field Dependence of the Relaxation Rate: The Berry Phase

The tunnel splitting can be easily evaluated, provided that the parameters of the spin Hamiltonian are known, and for the ± 10 pair it has been calculated to be ca. 10^{-11} K for $Mn_{12}ac$ and ca. 10^{-8} K for Fe_8Br. The tunneling is however possible only if the longitudinal field experienced by the spin of the cluster is small enough to keep the degeneracy of the two states within the tunnel splitting. This corresponds to an extremely narrow field range, of the order of 10^{-9} T for Fe_8Br, which would make the tunneling very hard to be observed. We will come back to this subject in the next section. The tunnel splitting can be, however, easily measured if after having saturated the magnetization a longitudinal field oscillating around the resonance value is applied so that the strict resonance condition is satisfied [64]. At each passage over the resonance a certain amount of the magnetization reverses its direction according to the Landau-Zener model [65]. The probability P of this process (see Fig. 4.4) is given by:

$$P_{M,M'} = 1 - \exp\left[\frac{\pi \Delta^2_{M,M'}}{2\hbar g\mu_B |M - M'|(\mu_0 \,\mathrm{d}H/\,\mathrm{d}t)}\right], \qquad (4.7)$$

where M and M' are the quantum numbers of the state involved in the resonance transition, g is the Lande factor close to 2 and $\mathrm{d}H/\mathrm{d}t$ is the field sweeping rate.

If this experiment, monitoring the ± 10 quantum transition, is repeated applying a transverse field, the tunnel splitting is expected to increase as the terms not commuting with $\boldsymbol{S_z}$ are known to promote the tunneling. These experiments have been performed on Fe_8Br using the μ-SQUID technique [66]. The static transverse field has been applied at different azimuth angles in

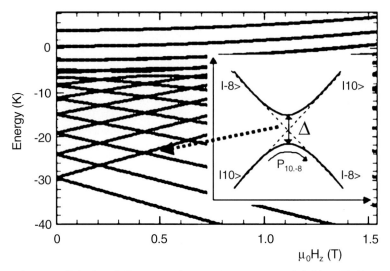

Fig. 4.4. Calculated Zeeman splitting due to an axial field in Fe_8Br with the parameters of ref. [25]. In the inset a schematic representation of an avoided crossing between the $M = 10$ and $M = -8$ states where the Landau-Zener transition probability P is indicated

the xy plane. The expected increase of the tunneling splitting is observed (see Fig. 4.5) when the static field is parallel to the intermediate y axis. As the direction of the transverse field approaches the hard axis oscillations of the tunnel splittings become visible. Several deep minima almost regularly spaced are observed for measurements along the hard axis, as shown in Fig. 4.5. This behavior has been predicted but never observed before [67]. It has been attributed to topological interference between two degenerate

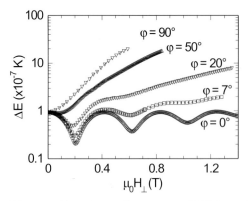

Fig. 4.5. Experimental transverse field dependence of the tunnel splitting in Fe_8Br obtained through the Landau-Zener model for different azimuthal angle ϕ, where $\phi = 0°$ corresponds to the hard axis of magnetization

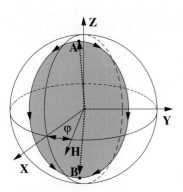

Fig. 4.6. Unit sphere showing degenerate minima A and B which are joined by two tunnel paths of opposite chirality. The hard, medium, and easy axes are in the x, y and z direction, respectively. The transverse field is applied in the xy plane at an azimuth angle ϕ. At zero longitudinal applied field, the giant spin reversal results from the interference of two quantum spin paths of opposite direction parallel to the easy anisotropy plane yz. Destructive interference, that is a quench of the tunneling rate, occurs whenever the shaded area is $l\pi S$, for l odd

tunnel pathways. In fact in a semi-classical picture the magnetization can be described as a vector pointing towards the easy direction, $+z$ in Fig. 4.6, if the system has been positively magnetized, but an equivalent minimum is observed along the $-z$ direction. If a transverse field is applied, the two degenerate minima are now described by the A and B points of Fig. 4.6. The reversal of the magnetization corresponds to a transition from A to B which can occur through two degenerate pathways of opposite chirality in the xy plane. Garg suggested that they lead to a destructive interference with quenching of the tunnel splitting when the area comprised by the two pathways on the unitary sphere is equal to $l\pi/S$ [67], being l an odd integer and S the spin value. For a second order bi-axial anisotropy the periodicity of the quenching is given by [67–69]:

$$\Delta H = \frac{2k_B}{g\mu_B}\sqrt{2E(E+D)}, \tag{4.8}$$

where D and E have been defined in (4.2).

It is evident that when the field is applied in the yz plane, the tunnel pathway is unique and interference effects are indeed not experimentally observed in this case as shown in Fig. 4.5. This effect has some similarity to the topological phase, the so-called Berry phase [70], observed for instance in superconductors.

In a quantum picture, more appropriate for the relatively small spin value of molecular clusters, the observed oscillation can be easily reproduced by diagonalizing the $(2S+1) \times (2S+1)$ hamiltonian matrix. The oscillatory behavior of the tunnel splitting of Fe_8Br is qualitatively reproduced by using the spin hamiltonian parameters determined by HFEPR experiments [25], but the periodicity is 0.32 T instead of 0.40 T as experimentally observed. It is however sufficient to multiply by four the forth order transverse term to correctly reproduce the periodicity. It is important to stress here that the tunnel splitting within the lowest states are strongly affected by the higher order terms of the transverse anisotropy to which HFEPR or inelastic neutron scattering experiments are not very sensitive.

The Landau-Zener experiments allow also to measure the tunnel splitting between any pair of levels which are brought in energy coincidence by oscillating the longitudinal field around the selected resonance, as shown in Fig. 4.4. For instance the longitudinal field can oscillate around a small negative value that brings the -10 and $+9$ states at the same energy. If the experiment is repeated for different values of the static field applied along the hard axis, again an oscillatory behavior is observed. This has however the opposite phase compared to the $+10 \to -10$ transition. The $-10 \to +8$ transition has instead the same phase, as can be seen in Fig. 4.7. This is a nice evidence of the parity effect originated by the symmetry of the spin hamiltonian. Numerical simulations nicely reproduce this parity effect [66]. Studies of the temperature dependence of the Landau Zener transition rates gives access to the topological quantum interference between excited spin levels [71]. Such measurements allowed also to estimate the life time of the excited energy levels and are in agreement with calculations [72].

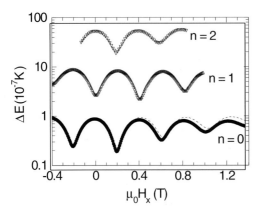

Fig. 4.7. Experimental transverse field dependence of the tunnel splitting in Fe_8Br obtained through the Landau-Zener model when oscillating the longitudinal field around the S and $(-S + n)$ transition for $n = 0$ (zero longitudinal field), $n = 1$, and $n = 2$. The oscillations relative to the transition with $n = 1$, corresponding to $M = 10$ and $M' = -9$, have the opposite phase

The experiments on Fe_8Br we have described here have revealed for the first time a topological effect on the magnetization and have provided a further confirmation of the tunneling picture used to justify the dynamics of the magnetization of molecular clusters. They might have also a technological interest. In fact if the hysteresis loop along the easy axis is measured by applying also a static transverse field, the loop is strongly modified. For instance in Fig. 4.8 we see that the step observed at $H_z = 0.2\,\mathrm{T}$ is strongly enhanced by the application of a transverse field $H_x = 0.19\,\mathrm{T}$ while it remains almost unaltered if a stronger field $H_x = 0.38\,\mathrm{T}$ is applied. The topological interference effects give therefore the possibility of controlling the coercivity

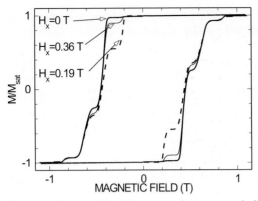

Fig. 4.8. Longitudinal hysteresis loops recorded for a single crystal of Fe_8Br in the pure tunneling regime with three values of a static field applied along the hard axis. The step at ca. 0.2 T is sensibly enhanced for $H_x = 0.19$ T

through the application of a transverse field. This is indeed a very hot point in the design of novel materials for high density data storage. In fact a higher density requires a higher coercivity to increase the stability of the data, but at the same time higher fields are required for the writing process. The coercivity of molecular clusters is strongly reduced by the application of an appropriate transversal field.

4.6 The Role of Dipolar Fields: The Non-Exponential Relaxation

Below 360 mK the Fe_8Br molecular clusters display a clear crossover from thermally activated relaxation to a temperature independent quantum regime, with a pronounced resonance structure of the relaxation time as a function of the external field. It was surprising however that the observed relaxation of the magnetization in the quantum regime was found to be non-exponential and the resonance width orders of magnitude too large compared to the tunnel splitting [63,73]. Furthermore we have already shown that the energy bias due to local field must be smaller than 10^{-9} T to allow tunneling, while the typical intermolecular dipole fields are of the order of 0.05 T. It seems therefore that almost all molecules should be blocked from tunneling by a very large energy bias. Prokof'ev and Stamp [74] have suggested a solution to this dilemma by proposing that fast dynamic nuclear fluctuations broaden the resonance, and the gradual adjustment of the dipole fields in the sample caused by the tunneling, brings other molecules into resonance thus allowing continuous relaxation. We do not review this theory [75] but focus on one particular application which is interesting for molecular clusters [76]. Prokof'ev and Stamp showed that at a given longitudinal applied field H_z, the magnetization of a crystal of molecular clusters should relax at short times with a square-root time dependence which is due to a gradual adjustment of the dipole fields in the sample caused by the tunneling:

$$M(H_z, t) = M_{in} + (M_{eq}(H_z) - M_{in})\sqrt{\Gamma_{sqr}(H_z)t} \ . \tag{4.9}$$

Here M_{in} is the initial magnetization at time $t = 0$ (after a rapid field change) and $M_{eq}(H_z)$ is the equilibrium magnetization at H_z. The rate function $\Gamma_{sqr}(H_z)$ is proportional to the normalized distribution $P(H_z)$ of molecules which are in resonance at H_z:

$$\Gamma_{sqr}(H_z) = a \frac{\xi_0}{E_d} \frac{\Delta_{\pm s}}{4\hbar} P(H_z) \ , \tag{4.10}$$

where ξ_0 is the linewidth contribution coming from the nuclear spins, E_D is the half-width of $P(H_z)$, and a is a constant of the order of unity which depends on the sample shape. If these simple relations are true, then measurements of the short time relaxation as a function of the applied field H_z gives directly the distribution $P(H_z)$, and allows one to measure the tunnel splitting $\Delta_{\pm S}$. Experimentally the square root time dependence of the magnetization is observed in Fe$_8$Br also by cooling in zero field and then applying a small field, as shown in Fig. 4.9.

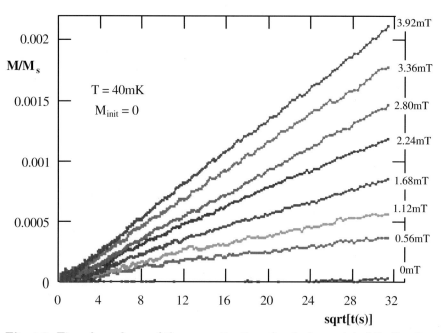

Fig. 4.9. Time dependence of the magnetization of a single crystal of Fe$_8$Br after cooling in zero field and then applying a longitudinal field. The short time (< 1000 s) data show a linear behavior if plotted vs $\cdot\sqrt{t}$

4 Quantum Tunneling of the Magnetization in Molecular Nanoclusters

Motivated by the Prokof'ev – Stamp theory, we developed a method, called the hole digging method, which allows us to study directly the influence of local field fluctuations [77]. In fact after a rapid field change, the resulting short time relaxation of the magnetization is directly related to the number of molecules, which are in resonance at the given applied field. Contrary to NMR, which probes the interaction between individual nuclear spins and the electron spins, our method probes the influence of all the nuclear spins, which are coupled to the giant spin of the molecular cluster.[1] The hole digging method consists of three steps as schematized in Fig. 4.10.

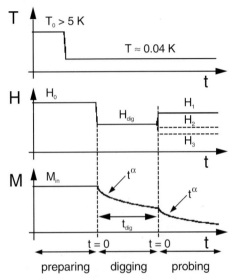

Fig. 4.10. Schematic representation of the temperature (*top*), applied field (*middle*), and magnetization (*bottom*) variation during the three steps of the hole-digging procedure. The temperature of the sample is rapidly quenched from high temperature to the investigate temperature (*top*). The magnetic field is kept at H_0 during the cooling process then H_{dig} is applied for a time period t_{dig} and after the measuring field H_1 is set. The procedure is repeated for different measuring field H_2, H_3, etc.

1) *Preparing the initial state*: A well defined initial magnetization state of the crystal of molecular clusters can be achieved by rapidly cooling the sample from high down to low temperatures in an applied field H_z. For zero applied field the demagnetized state, while for rather large applied fields, the saturated magnetization state of the entire crystal, respectively are reached. When the quench is fast (of the order of 1 s), the sample's magnetization has

---

[1] It is important to notice that the experimental method of the hole digging should work with any short time relaxation law, for example square root or power law.

not enough time to relax, either by thermal or quantum transitions. This procedure yields a frozen thermal equilibrium distribution.

2) *Modifying the initial state – hole digging*: After preparing the initial state, a field H_{dig} is applied during a time t_{dig}, called 'digging field' and 'digging time,' respectively. During the digging time and depending on H_{dig}, a fraction of the molecular spins tunnel and they reverse the direction of magnetization. The field sweeping rate to apply H_{dig} should be fast enough to minimize the change of the initial state during the field sweep.

3) *Probing the final state*: Finally, a field H_z is applied to measure the short time relaxation from which one observes $\Gamma_{\text{sqr}}(H_z)$, which is related to the number of spins free for tunneling after step (2). The entire procedure is then repeated many times, but at other fields H_z yielding $\Gamma_{\text{sqr}}(H_z, H_{\text{dig}}, t_{\text{dig}})$, which is related to the distribution of spins $P(H_z, H_{\text{dig}}, t_{\text{dig}})$ that are still free for tunneling after the hole digging. For $t_{\text{dig}} = 0$, this method maps out the initial distribution. In Fig. 4.11 the field dependence of the $\Gamma_{\text{sqr}}(H_z)$ for three different values of the initial magnetization are reported. The narrowest distribution is observed for the almost saturated sample, $M_{\text{in}} = -0.998 M_S$.

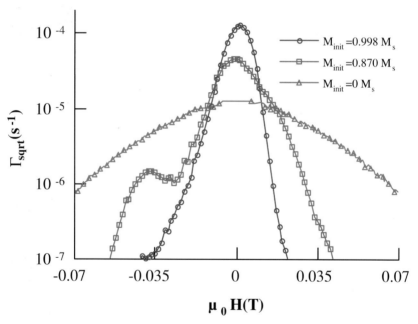

Fig. 4.11. Dependence of the short time tunneling rate as a function of the longitudinal field for three different initial states characterized by the initial magnetization almost saturated $M_{\text{init}} = -0.998 M_s$ (○), partially saturated $M_{\text{init}} = -0.870 M_s$ (□), and demagnetized state $M_{\text{init}} = 0$ (△). The structure observed for the partially saturated sample is due to those clusters which experience the dipolar environment generated by the reversal of one neighboring cluster along the a (-0.04 T), b (0.035 T), and c (0.025 T) crystallographic directions

A proof of the power of the method in providing information on the local bias field distribution is given by the remarkable structure seen for $M_{in} = -0.87 M_S$. The peak at -0.04 T as well as the shoulder at $+0.02$ T and $+0.04$ T are originated by the clusters which have one nearest neighbor cluster with reversed magnetization. The first signature corresponds to the reversal of the neighboring cluster along the a crystallographic axis, which almost coincides with the easy axis of magnetization z, while the other two are due to the clusters along b and c. The shoulder at 0.02 T corresponding to the reversal of the neighbor cluster along the crystallographic axes with the longest period and therefore at larger distance. These results are in good agreement with simulations [78,79].

The experimental results of Fig. 4.11 suggest that when a hole is dug in a saturated sample it is dominated by the change of intermolecular dipolar fields during the digging. However, in special conditions, we could reduce the change of intermolecular dipolar fields in order to be sensitive to local field fluctuations coming from nuclear spins. This can be done for instance by digging a hole into the tail of the dipolar distribution of a demagnetized sample as shown in Fig. 4.12. In this condition almost all molecules are out of resonance and only a very small fraction ($< 10^{-4}$) might be brought into resonance by the hyperfine field fluctuations. Therefore, for short digging times, a practically negligible variation of the intermolecular field occurs and the hole linewidth reflects directly by the hyperfine field fluctuations [80]

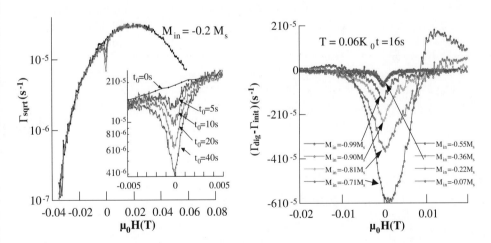

Fig. 4.12. On the left distribution of the tunneling rate in Fe_8Br after cooling in a longitudinal field so that $M = 0.2 M_s$. In the inset the hole obtained for digging time 0, 5, 10, 20, 40 s. On the right the hole-digging at 0.06 K for $t_{dig} = 16$ starting from different initial magnetization. For initial magnetization $M_{in} < 0.5 M_s$ the linewidth of the hole becomes constant showing that the line-width is not depending on the initial magnetization

(see next section). This is the intrinsic linewidth mentioned in the next section. An alternative way is that of digging the hole in a very diluted sample, so that intermolecular dipolar fields can be neglected, and the hole digging method reflects directly the hyperfine field fluctuations. We will see in the next section that this situation is achieved in $Mn_{12}ac$ sample where we used a very diluted second species in order to map out the hyperfine field fluctuations [81].

4.7 The Role of the Nuclear Magnetic Moments: The Isotope Effect

The role of the nuclear magnetic moment in promoting tunneling has been experimentally confirmed by isotopically modifying the Fe_8Br clusters [80]. In particular ^{57}Fe enriched starting material (95%) has been used in order to increase the hyperfine field. ^{57}Fe, which has a natural abundance of 2.6% the rest being the non-magnetic ^{56}Fe, has $I = 1/2$. In order to decrease the hyperfine field, a partial deuteration has been performed by using the protonated ligand *tacn* in deuterated solvents. In this way only the mobile protons (of the water, of the bridging OH^- and of the NH groups of the *tacn* ligand) are expected to be replaced by deuterons. ^2H has indeed a larger nuclear spin, $I = 1$ vs. $I = 1/2$ of the proton, but a significantly smaller gyromagnetic factor, thus reducing the nuclear magnetic moment.

The two isotopically substituted samples have been carefully checked by single crystal x-ray analysis and moreover their transverse field dependence of the tunneling rate has been checked according to the procedure of Sect. 4.5.

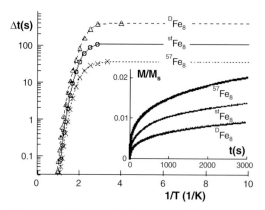

Fig. 4.13. Time needed to relax 1% of the saturation magnetization for the three isotopic Fe_8Br crystals vs the inverse of temperature. In the inset the time dependence of the magnetization of the three crystals measured after cooling in zero field and then applying a longitudinal field of 42 mT in the pure tunneling regime ($T = 40$ mK)

The position of the quenching of the tunneling rate remains unchanged in the three samples confirming that the magnetic anisotropy is not influenced by the isotopic enrichment. On the contrary the relaxation time of the magnetization is significantly changed. In Fig. 4.13 we report the time needed to relax 1% of the saturation magnetization. The ^{57}Fe$_8$ compound shows the fastest relaxation in agreement with the larger hyperfine field, where the slowest relaxation is observed for the deuterium enriched crystal. This marked isotopic effect does not depend on the mass, which is increased in both isotopically enriched samples, but rather on the nuclear magnetism.

The hole digging method previously described has allowed a comparison of the intrinsic linewidth of the tunneling resonance for the three isotopic samples. The conditions that minimize the dipolar broadening have been accurately selected so that the observed linewidth is essentially dominated by the hyperfine contribution. Such a condition corresponds to a longitudinal field of 42 mT and to a transverse field applied along the hard axis of 200 mT. The linewidth of the hole depends on the amount of magnetization that has been reversed ΔM_{dig}. By extrapolating the linewidth to $\Delta M_{\text{dig}} = 0$ it is possible to get the intrinsic linewidth. This results to be 0.6 ± 1 mT, 0.8 ± 1 mT and 1.2 ± 1 mT for the deuterated, natural and ^{57}Fe enriched samples as reported in Fig. 4.14.

In order to confirm the nuclear origin of the difference, the hyperfine field acting on the ground $S = 10$ state has been calculated. The hyperfine interaction between the total spin S of the cluster and the magnetic nuclei can be decomposed into the sum of terms related to the interactions between the magnetic moment I_i of the ith nucleus and the individual S_j spin, assumed

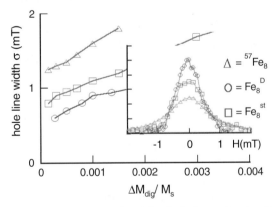

Fig. 4.14. Line-width of the hole burnt in the distribution of tunnel rate for the three isotopic crystals of Fe$_8$Br as a function of the fraction of magnetization reversed during the hole-digging procedure. In the inset the experimental hole lineshape. The experimental condition of $H_z = 42$ mT and $H_x = 200$ mT have been selected as those for which the narrowest hole is obtained

to be localized on the jth iron center:

$$H_{hf} = \sum_i S \cdot A_i \cdot I_i = S \cdot \sum_i (\sum_j C_j A_{ij}) \cdot I_i \, . \tag{4.11}$$

The projection coefficients C_j in (4.11) depend on the wavefunction of the ground $S = 10$ state, which can be calculated by diagonalizing the $S = 10$ block (6328 × 6328) of the exchange spin-Hamiltonian matrix of $H = \sum_{j \neq k} J_{jk} S_j \cdot S_k$. The largest contribution to the ground wavefunction is provided by the function in which the spin of Fe$_3$ and Fe$_4$ in Fig. 4.1 are antiparallel to the remaining ones, as recently confirmed by polarized neutron diffraction data [11]. The C_j coefficients can be calculated using recurrently the projection technique for coupled angular momenta. They are $C_3 = C_4 = -5/22$ for the spins pointing down and $C_1 = C_2 = C_5 = \ldots = C_8 = 8/33$ for the remaining iron spins. The single spin hyperfine coupling constant A_{ij} contains both through-space (dipolar) and through-bond (contact) contributions. While the first one could be easily calculated using the point-dipole approximation, the last one cannot easily be evaluated in such a complex system. *Ab-initio* calculation based on the Density Functional approach has been used to evaluate the contact term of the hyperfine coupling on the proton of the bridging OH groups on the nitrogen and proton of the NH group of the ligands by using a model dimeric species [(NH$_3$)$_4$Fe(μ − OH)$_2$Fe(NH$_3$)$_4$] [80]. The results have only an indicative value, but reveal that the contact term has the same order of magnitude as the dipolar one. A more quantitative analysis can be performed in the case of the contribution of the ^{57}Fe nuclei because in this case the contact term with the spin of the same iron atom is strongly dominating. Values of $A(^{57}\text{Fe})$ around 1 mT are reported [82] in the literature. The hyperfine interaction with 8 non equivalent $I = 1/2$ gives 2^8 states. This distribution of hyperfine fields calculated from (4.11) can be approximated by a Gaussian distribution whose linewidth $\sigma(^{57}\text{Fe})$ is ca. 0.9(1) mT. This extra contribution adds to the linewidth of the natural Fe$_8$Br cluster in a geometrical way:

$$\sigma(^{57}\text{Fe}_8) = \sqrt{[\sigma(\text{Fe}_8\text{Br})]^2 + [\sigma(^{57}\text{Fe})]^2} \tag{4.12}$$

where $\sigma(\text{Fe}_8\text{Br})$ is the experimental linewidth found for the natural Fe$_8$. Equation (4.12) gives for the linewidth an estimated $\sigma(^{57}\text{Fe}_8) = 1.1(1)$ mT in good agreement with the experimental value of 1.2(1) mT.

The role of the hyperfine field in promoting the tunneling in Fe$_8$Br has been clearly evidenced by the experiment mentioned above. Similar experiments have been performed on Mn$_{12}$ac; even if in this case the isotopic enrichment does not lead to a significant variation of the hyperfine field, as this is dominated by the contribution of ^{55}Mn, $I = 5/2$, which is the only stable isotope of manganese. Experiments performed in the temperature range 1.5–4 K, temperature at which the relaxation can be followed, failed to dig any hole in the distribution of the relaxation rate around a resonance as expected for a homogeneously broadened resonance line. The line-width is in fact given by the longitudinal field distribution of the transition probability for the below-barrier process occurring in the upper part of the barrier, as observed by Friedman et al. [83] and theoretically established by Loss et al. [60].

In the very rich series of dodecanuclear manganese clusters, where carboxylate ligands different from acetate can be used [7], it is often observed that at least two isomeric forms can be obtained [5,84]. The two isomeric forms are characterized by a modification of the coordination geometry around a manganese(III) center, with the elongation axis typical of the Jahn-Teller distorsion of manganese(III) pointing toward a different direction. As we have

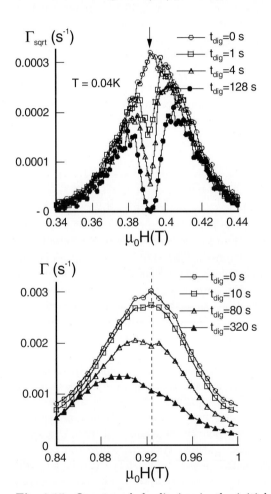

Fig. 4.15. Quantum hole digging in the initial dipolar distribution of tunneling rate for the fast relaxing impurities in $Mn_{12}ac$ (*top*). Same experiments performed at $T = 2$ K for the majority species of $Mn_{12}ac$

seen in Sect. 4.2 the magnetic anisotropy of the spin ground state is related to the single ion contribution and the two isomers can differ significantly in the magnetic anisotropy and consequently in the dynamics of the magnetization. μ-SQUIDs measurements have been performed on a small single crystal of $Mn_{12}ac$ and have revealed the existence of a fast relaxing species whose distribution is homogeneous in the crystal [81]. The amount of these defect sites is slightly variable from crystal to crystal and their magnetization accounts for ca. 5% of the saturation magnetization. The fast relaxing species can be easily investigated in low fields and at temperatures below 1.5 K, as the magnetization of the major one is practically frozen at these temperatures. The stepped hysteresis cycle of the defective sites revealed a significantly smaller axial anisotropy, as the steps occurs at ca. 0.39 T instead of 0.45 T for the main species. The relaxation rate becomes temperature independent below 0.6 K and in the pure tunneling regime it is possible to dig a hole in the distribution of the tunneling rate. The linewidth of the hole is ca. 12 mT, as can be seen in Fig. 4.15, and is significantly larger than what has been observed in Fe_8Br in agreement with the larger hyperfine field generated by the ^{55}Mn nuclei. The hyperfine linewidth is however too small to detect a hole digging in the major species, where the homogenous broadening [60,83,85] due to the spin-phonon coupling is around 30 mT at $T > 1.5$ K (see Fig. 4.15).

4.8 Conclusions

Experiments and theoretical models have provided an accurate description of the low temperature dynamics of the magnetization of molecular clusters. Quantum effects have been evidenced and have provided unprecedented phenomena, like the quenching of the tunneling for topological interference or the magnetic isotope effect. What remains still debated is the possibility of observing quantum coherence between states of opposite magnetization [86]. Dipole-dipole and hyperfine interactions described in the last two sessions are source of decoherence. In other words when a spin has tunneled through the barrier, it experiences a huge modification of its environment (hyperfine and dipolar) which prohibits the back tunneling. Prokov'ev and Stamp [87] suggested three possible strategies to suppress the decoherence. (i) Choosing a system were the NMR frequencies far exceed the tunnel frequencies making any coupling impossible. (ii) Isotopically purifying the sample to remove all nuclear spins. (iii) Applying a transverse field to increase the tunnel rate to frequencies much larger than hyperfine field fluctuations. All three strategies are difficult to realize. However, some authors tried to realize the last one by performing EPR experiments in the presence of a large transverse field [88,89]. Absorption of radio-frequency electromagnetic fields were observed which might be due to induced transitions near the tunnel splitting. However, no experiments showed the oscillatory behavior in the time domain which might be evidenced by a spin echo type of experiment.

Concerning the perspectives of the field of single molecule magnets we expect that chemistry is going to play a major role through the synthesis of novel larger spin clusters with strong anisotropy. Effecting an increase in the magnitude of the blocking temperature is in fact a necessary condition for any potential application. We are actually involved in the attempts of depositing of the clusters on substrates (metallic or insulating), while some success has already been achieved in obtaining Langmuir-Blodgett films [90]. This is a necessary step forward in the utilization of the clusters as molecular memory devices. In fact one of the main problems for their use in applications is how to address a single molecule in the reading-writing process. The hole digging technique described here is providing a tool to select some molecules in a wide distribution, in analogy to what has been found in single-molecule spectroscopy [91,92]. Very recently molecular clusters have been proposed for an improvement of the Grover's algorithm [94] in quantum computing [93]. In conclusion we can say that the field of single molecule magnets is still young.

Acknowledgements. The financial support of the European Community through the 3MD EU Network contr. N° ERBFMRX-CT98 0181 and of Italian MURST is acknowledged.

References

1. D. Gatteschi, A. Caneschi, L. Pardi and R. Sessoli, Science **265**, 1054 (1994).
2. R. Sessoli, D. Gatteschi, A. Caneschi and M.A. Novak, Nature (London) **365**, 141 (1993).
3. B. Barbara and L. Gunther, Physics World **12**, 35 (1999).
4. A. Cornia, M. Affronte, A.G.M. Jansen, G.L. Abbati and D. Gatteschi, Angew. Chem., Int. Ed. Engl. **38**, 2264 (1999).
5. Z.M. Sun, D. Ruiz, E. Rumberger, C.D. Incarvito, K. Folting, A.L. Rheingold, G. Christou and D.N. Hendrickson, Inorg. Chem. **37**, 4758 (1998).
6. E.K. Brechin, J. Yoo, M. Nakano, J.C. Huffman, D.N. Hendrickson and G. Christou, J. Chem. Soc., Chem. Commun. 783 (1999).
7. G. Aromi, S.M.J. Aubin, M.A. Bolcar, G. Christou, H.J. Eppley, K. Folting, D.N. Hendrickson, J.C. Huffman, R.C. Squire, H.L. Tsai et al., Polyhedron **17**, 3005 (1998).
8. T. Lis, Acta Cryst. B **36**, 2042 (1980).
9. K. Wieghardt, K. Pohl, I. Jibril and G. Huttner, Angew. Chem., Int. Ed. Engl. **23**, 77 (1984).
10. J. Larionova, M. Gross, M. Pilkington, H. Andres, H. Stoeckli-Evans, H.U. Gudel and S. Decurtins, Angew. Chem. Int. Ed. Engl. **39**, 1605 (2000).
11. Y. Pontillon, A. Caneschi, D. Gatteschi, R. Sessoli, E. Ressouche, J. Schweizer and E. Lelievre-Berna, J. Am. Chem. Soc. **121**, 5342 (1999).
12. R.A. Robinson, P.J. Brown, D.N. Argyriou, D.N. Hendrickson and S.M.J. Aubin, J. Phys.: Condens. Matter **12**, 2805 (2000).
13. H.J. Eppley, H.-L. Tsai, N. de Vries, K. Folting, G. Christou and D.N. Hendrickson, J. Am. Chem. Soc. **117**, 301 (1995).

14. A.M. Gomes, M.A. Novak, W. Wernsdorfer, R. Sessoli, L. Sorace and D. Gatteschi, J. Appl. Phys. **87**, 6004 (2000).
15. K. Takeda and K. Awaga, Phys. Rev. B **56**, 14560 (1997).
16. A.L. Barra, F. Bencini, A. Caneschi, D. Gatteschi, C. Paulsen, C. Sangregorio, R. Sessoli, L. Sorace, Chem. Phys. Chem. **2**, 253 (2001).
17. A.L. Barra, A. Caneschi, A. Cornia, F.F. De Biani, D. Gatteschi, C. Sangregorio, R. Sessoli and L. Sorace, J. Am. Chem. Soc. **121**, 5302 (1999).
18. A. Abragam and B. Bleaney, *Electron Paramagnetic Resonance of Tansition Ions* (Dover, New York, 1986).
19. T. Moriya, G.T. Rado and H. Suhl, *Magnetism*, (Academic Press, New York, 1963), Vol. 1.
20. A. Bencini and D. Gatteschi, *EPR of Exchange Coupled Systems* (Springer-Verlag, Berlin, 1990).
21. G.L. Abbati, L.C. Brunel, H. Casalta, A. Cornia, A.C. Fabretti, D. Gatteschi, A.K. Hassan, A.G.M. Jansen, A.L. Maniero, L.A. Pardi et al., Chemistry Eur. J. **7**, 1796 (2001).
22. D. Gatteschi and R. Sessoli, *Magnetoscience – From Molecules to Materials*, J.S. Miller, M. Drillon, eds. (Wiley-VCH, Weinheim, 2002).
23. A. Cornia, D. Gatteschi and R. Sessoli, Coord. Chem. Rev. **219**, 573 (2001).
24. I. Chiorescu, R. Giraud, A.G.M. Jansen, A. Caneschi and B. Barbara, Phys. Rev. Lett. **85**, 4807 (2000).
25. A.L. Barra, D. Gatteschi and R. Sessoli, Chemistry Eur. J. **6**, 1608 (2000).
26. J.R. Friedman, Phys. Rev. B **57**, 10291 (1998).
27. A. Caneschi, D. Gatteschi, R. Sessoli, A.-L. Barra, L.C. Brunel and M. Guillot, J. Am. Chem. Soc. **113**, 5873 (1991).
28. A.L. Barra, P. Debrunner, D. Gatteschi, Ch.E. Schulz and R. Sessoli, Europhys. Lett. **35**, 133 (1996).
29. A.L. Barra, D. Gatteschi and R. Sessoli, Phys. Rev. B **56**, 8192 (1997).
30. S. Hill, J.A.A.J. Perenboom, N.S. Dalal, T. Hathaway, T. Stalcup and J.S. Brooks, Phys. Rev. Lett. **80**, 2453 (1998).
31. A.A. Mukhin, V.D. Travkin, A.K. Zvezdin, A. Caneschi, D. Gatteschi and R. Sessoli, Physica B **284**, 1221 (2000).
32. A. Cornia, M. Affronte, A.G.M. Jansen, D. Gatteschi, A. Caneschi and R. Sessoli, Chem. Phys. Lett. **322**, 477 (2000).
33. M. Hennion, L. Pardi, I. Mirebeau, E. Suard, R. Sessoli and A. Caneschi, Phys. Rev. B **56**, 8819 (1997).
34. Y.C. Zhong, M.P. Sarachik, J.R. Friedman, R.A. Robinson, T.M. Kelley, H. Nakotte, A.C. Christianson, F. Trouw, S.M.J. Aubin and D.N. Hendrickson, J. Appl. Phys. **85**, 5636 (1999).
35. I. Mirebeau, M. Hennion, H. Casalta, H. Andres, H.U. Gudel, A.V. Irodova and A. Caneschi, Phys. Rev. Lett. **83**, 628 (1999).
36. G. Amoretti, R. Caciuffo, J. Combet, A. Murani and A. Caneschi, Phys. Rev.B **62**, 3022 (2000).
37. R. Caciuffo, G. Amoretti, A. Murani, R. Sessoli, A. Caneschi and D. Gatteschi, Phys. Rev. Lett. **81**, 4744 (1998).
38. A.L. Barra, D. Gatteschi, R. Sessoli, G.L. Abbati, A. Cornia, A.C. Fabretti and M.G. Uytterhoeven, Angew. Chem., Int. Ed. Engl. **36**, 2329 (1997).
39. M.R. Pederson and Khanna S. N: Phys. Rev. B **60**, 9566 (1999).
40. M.R. Pederson, D.V. Porezag, J. Kortus and S.N. Khanna, J. Appl. Phys. **87**, 5487 (2000).

41. A.H. Morrish, *The Physical Principles of Magnetism* (John Wiley & Sons, Inc., New York, 1966).
42. M.A. Novak and R. Sessoli, *Quantum Tunneling of Magnetization – QTM'94* (Kluwer Academic Publishers, 1995), pp. 171–188.
43. J. Villain, F. Hartman-Boutron, R. Sessoli and A. Rettori, Europhys. Lett. **27**, 159 (1994).
44. F. Hartmann-Boutron, P. Politi and J. Villain, Int. J. Mod. Phys. B **10**, 2577 (1996).
45. J.F. Fernandez and J.J. Alonso , Phys. Rev. B. **62**, 53 (2000).
46. A.M. Gomes, M.A. Novak, R. Sessoli, A. Caneschi and D. Gatteschi, Phys. Rev. B **57**, 5021 (1998).
47. R. Sessoli, Molecular Crystals Liquid Crystals **274**, 145 (1995).
48. A. Caneschi, T. Ohm, C. Paulsen, D. Rovai, C. Sangregorio and R. Sessoli, J. Magn. Magn. Mater. **177**, 1330 (1998).
49. M.R. Cheesman, V.S. Oganesyan, R. Sessoli, D. Gatteschi and A.J. Thomson, Chem. Commun., 1677 (1997).
50. A. Fort, A. Rettori, J. Villain, D. Gatteschi and R. Sessoli, Phys. Rev. Lett. **80**, 612 (1998).
51. J.R. Friedman, M.P. Sarachik, J. Tejada and R. Ziolo, Phys. Rev. Lett. **76**, 3830 (1996).
52. L. Thomas, F. Lionti, R. Ballou, D. Gatteschi, R. Sessoli and B. Barbara, Nature (London) **383**, 145 (1996).
53. A.D. Kent, Y.C. Zhong, L. Bokacheva, D. Ruiz, D.N. Hendrickson and M.P. Sarachik, Europhys. Lett. **49**, 521 (2000).
54. E.M. Chudnovsky and D.A. Garanin, Phys. Rev. Lett. **79**, 4469 (1997).
55. D.A. Garanin, X.M. Hidalgo and E.M. Chudnovsky, Phys. Rev. B **57**, 13639 (1998).
56. L. Bokacheva, A.D. Kent and M.A. Walters, Phys. Rev. Lett. **85**, 4803 (2000).
57. F. Fominaya, T. Fournier, P. Gandit and J. Chaussy, Rev. Sci. Instrum. **68**, 4191 (1997).
58. Z.H. Jang, A. Lascialfari, F. Borsa and D. Gatteschi, Phys. Rev. Lett. **84**, 2977 (2000).
59. F. Luis, J. Bartolomé and J. F. Fernàndez, Phys. Rev. B **57**, 505 (1998).
60. M.N. Leuenberger and D. Loss, Europhys. Lett. **46**, 692 (1999).
61. M.N. Leuenberger and D. Loss, Phys. Rev.B **61**, 1286 (2000).
62. A. Caneschi, D. Gatteschi, C. Sangregorio, R. Sessoli, L. Sorace, A. Cornia, M.A. Novak, C. Paulsen and W. Wernsdorfer, J. Magn. Magn. Mater. **200**, 182 (1999).
63. C. Sangregorio, T. Ohm, C. Paulsen, R. Sessoli and D. Gatteschi, Phys. Rev. Lett. **78**, 4645 (1997).
64. L. Gunther, Europhys. Lett. **39**, 1 (1997).
65. C. Zener, Proc. R. Soc. London A **137**, 3237 (1932).
66. W. Wernsdorfer and R. Sessoli, Science **284**, 133 (1999).
67. A. Garg, Europhys. Lett. **22**, 205 (1993).
68. A. Garg, Physica B **280**, 269 (2000).
69. J. Villain and A. Fort, Eur. Phys. J. **17**, 69 (2000).
70. M.V. Berry, Proc. Roy. Soc. London A **392**, 45 (1984).
71. W. Wernsdorfer, R. Sessoli, A. Caneschi, D. Gatteschi and A. Cornia, Europhys. Lett. **50**, 552 (2000).

72. M.N. Leuenberger and D. Loss, Phys. Rev. B **61**, 12200 (2000).
73. T. Ohm, C. Sangregorio and C. Paulsen, J. Low Temp. Phys. **113**, 1141 (1998).
74. N.V. Prokof'ev and P.C.E. Stamp, Phys. Rev. Lett. **80**, 5794 (1998).
75. N.V. Prokof'ev and P.C.E. Stamp, Rep. Prog. Phys. **63**, 669 (2000).
76. N. V. Prokof'ev and P.C.E. Stamp, J. Low Temp. Phys. **113**, 1147 (1998).
77. W. Wernsdorfer, T. Ohm, C. Sangregorio, R. Sessoli, D. Mailly and C. Paulsen, Phys. Rev. Lett. **82**, 3903 (1999).
78. T. Ohm, C. Sangregorio and C. Paulsen, Eur. Phys. J. B **6**, 195 (1998).
79. A. Cuccoli, A. Fort, A. Rettori, E. Adam and J. Villain, Eur. Phys. J. B **12**, 39 (1999).
80. W. Wernsdorfer, A. Caneschi, R. Sessoli, D. Gatteschi, A. Cornia, V. Villar and C. Paulsen, Phys. Rev. Lett. **84**, 2965 (2000).
81. W. Wernsdorfer, R. Sessoli and D. Gatteschi, Europhys. Lett. **47**, 254 (1999).
82. B.R. McGarvey, Transition Metal Chemistry **3**, 89 (1966).
83. J.R. Friedman, M.P. Sarachik and R. Ziolo, Phys. Rev. B **58**, R14729–R14732 (1998).
84. S.M.J. Aubin, Z.M. Sun, I.A. Guzei, A.L. Rheingold, G. Christou and D.N. Hendrickson, J. Chem. Soc., Chem. Commun. 2239 (1997).
85. D.A. Garanin, E.M. Chudnovsky and R. Schilling, Phys. Rev.B **61**, 12204 (2000).
86. A.J. Leggett in *Quantum Tunneling of Magnetization – QTM'94* (Kluwer Academic Publishers, 1995), p. 1.
87. N.V. Prokof'ev and P.C.E. Stamp, *Quantum Tunneling of Magnetization – QTM'94*, (Kluwer Academic Publishers, 1995), p. 369.
88. E. Del Barco, J.M. Hernandez, J. Tejada, N. Biskup, R. Achey, I. Rutel, N. Dalal and J. Brooks, Phys. Rev.B **62**, 3018 (2000).
89. E. Del Barco, N. Vernier, J.M. Hernandez, J. Tejada, E.M. Chudnovsky, E. Molins and G. Bellessa, Europhys. Lett. **47**, 722 (1999).
90. M. Clemente-Leon, H. Soyer, E. Coronado, C. Mingotaud, C.J. Gomez-Garcia and P. Delhaes, Angew. Chem., Int. Ed. Engl. **37**, 2842 (1998).
91. X.S. Xie and J.K. Trautman, Annu. Rev. Phys. Chem. **49**, 441 (1998).
92. S.J. Zilker, J. Friebel, D. Haarer, Y.G. Vainer and R.I. Personov, Chem. Phys. Lett. **289**, 553 (1998).
93. M.N. Leuenberger and D. Loss, Nature **410**, 789 (2001).
94. L.K. Grover, Science **280**, 228 (1998).

5 Magnetism of Free and Supported Metal Clusters

J.P. Bucher

Stern-Gerlach experiments on beams of metal clusters provide detailed information on their intrinsic magnetic moments but also on the relaxation processes involved when the free clusters cross a magnetic field. For transition metal clusters of a few 10 to a few 100 atoms the observable projection of the magnetic moment onto the field axis (similar to magnetization) measured from the clusters deflection, scales with magnetic field, clusters size and inverse vibrational temperature. The measurements are in quantitative agreement with a picture in which the cluster moments are subject to rapid orientational fluctuations. Intrinsic magnetic moments per atom in excess of the bulk values are obtained for transition metal clusters such as Co, Ni, and Fe, while rare earth clusters such as Gd or Tb possess a lower magnetic moment than the bulk. For the smallest clusters, the intrinsic magnetic moment usually follows an oscillatory behavior as a function of size. In the case of rare earth clusters the magnetic behavior may even change its character from one cluster size to another. Except for some "magic numbers", for which the statistical interpretation still holds, an anomalous spreading of the deflection profile is observed. This spreading is due to a strong coupling of the magnetic moment with the cluster body. When the moment is locked to the lattice by strong crystal field anisotropies, the rotational temperature starts to play an important role in the interpretation of experimental data. This distinct behavior points to the fact that 3d and 3f ferromagnetism react quite differently to a confined geometry. This dissimilarity is due in part to a different relative importance of magnetic anisotropy energy and exchange energy.

One of the most fascinating discoveries in surface physics during the last decade was that diffusing atoms on a surface may lead to well organized aggregates whose structure can be manipulated experimentally. In other words, the density, shape, distribution of clusters (or any small entity of atoms) can be controlled to a large extent without any external intervention by controlling the physical parameters such as substrate temperature, substrate crystallographic symmetry and deposition rate. When the small entities bear a net, giant magnetic moment, this view considerably helps understanding the critical magnetic phenomena involved during growth and coalescence of the deposits. Furthermore, on some systems, the clusters spontaneously organize on the surface at the nanometer scale into a regular periodic network of islands. Magnetic phase transitions can then be studied in a controlled fashion by adjusting the dimensionality of the systems from 0D to 1D and finally 2D. These concepts will be illustrated on examples from the recent literature. New developments and applications will be emphasized.

5.1 Introduction

Small metal clusters have become the center of interest in various interdisciplinary subjects such as catalysis, macroscopic quantum tunneling and Coulomb blockade devices [1–4]. Clusters can be studied on their own for example in molecular beams [5], but they also constitute the ultimate state of integration in electronic and optoelectronic devices and as such they are the subject of many interesting studies in the form of supported particles.

Physical and chemical properties of clusters often change dramatically as a function of cluster size, temperature and vectorial fields [6]. The magnetic properties of metal clusters are among the most fascinating and less well understood. Most atoms lose their magnetic moments when they condense into the solid phase, but there are three classes of elements, the 3d transition metals, the 4f, or rare earth series, and the 5f or actinide series which retain their magnetism in the condensed phase. Much of our understanding of solid state magnetism is drawn from the former two groups. In the past years, the study of magnetic properties of transition metal and rare earth clusters has developed into a new and attractive field of research [7–13].

Magnetic materials and devices made of well defined nanoscale particles form an important part of recent progress in spin electronics, magnetic data storage and sensors for giant magnetoresistance applications [14,15]. It is generally believed that each particle can store information corresponding to 1 bit. In this context, the controlled formation of ordered metal nanostructures on solid surfaces by self-organized growth [16–18], allows to anticipate new data storage technologies based on nanoscale dots with tunable densities in excess of 1 Tbits/in^2.

Although the beam experiments have been important in their own right in disentangling intrinsic magnetic properties of isolated non-interacting clusters, real systems that are technologically relevant, rely on clusters supported on substrates or embedded in matrices. These entities then interact with the substrate and mostly with each others. This review article presents two complementary aspects of cluster magnetism: (i) the fundamental properties of free, isolated clusters in the vacuum as they are measured in cluster beams, and (ii) the magnetic properties of cluster assemblies that form spontaneously on surfaces during deposition of metal vapor in a ultrahigh vacuum (UHV) environment.

The experimental technique that will be referred to for the discussion of free cluster magnetism is similar to the one that led Stern and Gerlach to discover space quantization in atoms [19]. It mainly involves the passage of clusters through an inhomogeneous magnetic field. These recent developments in molecular beam technology offer the unique opportunity to study clusters of sizes ranging from 2 to 500 atoms. The evolution of their magnetic properties can be followed very precisely, free from interaction with any matrix or support, by stepwise addition of atoms to a basic unit. In this regime, the atomic character becomes apparent and effects due to the surface, the low dimensionality [20] and symmetry [21] start to play important roles. Couplings between the dynamical parameters of the system must also be considered. The comparative study of magnetic properties of transition metal and rare earth clusters presents several advantages. Among others, it allows one to analyze the impact of a constrained geometry on the two distinct, well known classes of metallic magnetisms. It is that approach that is adopted in this review.

After discussing basic and experimental aspects of magnetic cluster beams, we review recent results obtained in the study of magnetic properties of transition metal and rare earth clusters. We discuss the validity of the hypotheses and theories that have been put forward to explain the available experimental data. In particular, the central role played by thermal fluctuations is analyzed. While the coupling between the magnetic moment and the lattice through anisotropies only provides a relaxation channel in the case of transition metal clusters, it leads to a nearly rigid coupling of the magnetic moment to the lattice in the case of rare earth clusters.

The synthesis and magnetic properties of organized metal hetero-structures on surfaces is reviewed herein, and examples are presented from recent research. We do not address the topic of magnetic particles in matrices which is a subject in itself and has its own specificity in the field of composite materials. Nanosize islands on metal surfaces spontaneously form as a result of condensation of metal vapor from the gas phase and subsequent nucleation and growth. By controlling the growth kinetics, nanostructures with particular properties can be synthesized [22,23]. Islanding on surfaces is also the obliged pathway for the growth of thin films. Therefore, each time it is possible, reference will be made to ultrathin magnetic films for which a large bulk of information exists [24]. Unfortunately, in most of the existing work, the initial stage of growth is only marginally addressed although it is central for the understanding of the magnetic properties of assembled structures on surfaces. There are a few exceptions however; see for example [25–27].

The possibility of getting self-organized islands on surfaces is addressed and the driving mechanisms is discussed. A large part of Sect. 5.6, however, is devoted to surface reconstruction and strain relaxation patterns owing to their central role as atomic scale templates for the organization of small entities at the nanometer scale. The magnetic properties of low dimensional systems, from isolated islands to the 2D limit, is revised in Sect. 5.7. The goal is to develop a good understanding of nanostructured magnetic materials in terms of their basic parameters such as grain size and density. Ultimately, magnetism is viewed as the result of interacting building blocks (spin blocks); a particularly fruitful approach in the case of self-organized cobalt dots. In this part of the work, we show how the interaction between dots develops towards the formation of magnetic domain structures.

5.2 Simple Considerations

5.2.1 Common Ideas on Magnetism

The origin of magnetic ordering in the solid state has been understood in principle since the work of Heisenberg and Dirac [28,29]. The exchange correlation of electrons of parallel spin holds them apart and hence reduces their Coulomb potential energy. This energy reduction does not occur if spins are antiparallel; therefore a pair of electrons will tend to have their spins aligned. Earlier theories of ferromagnetism have all been based on the Heisenberg model, in which it is assumed that a magnetic moment is localized on each individual atom and that they interact through an effective exchange parameter J. The assumption of localized moments is also built into the Weiss molecular field theory, whether we call the force that causes parallel spin alignment a molecular field or an exchange force. Other theories have been considered. In the band model or collective electron model, ferromagnetism appears as a result of spin imbalance in the 3d band. Since the 3d band is overlapped in energy by a much wider 4s band, this theory predicts correctly the non-integral value of the effective number of magnetic carriers [30,31]. More sophisticated theories have been developed that all rely on either one or both of these approaches [32]. These diverging views of magnetism have led to large debates and there is a rich literature on the subject [33]. Therefore only some aspects relevant to clusters is discussed in this brief introduction.

The knowledge that exchange forces are responsible for ferromagnetism has led to many semi-quantitative conclusions of great value. For example, it allows one to rationalize the appearance of ferromagnetism in some metals and not in others. Variations of the exchange integral with interatomic distance can be predicted. When the atoms are far apart, the 3d orbitals of transition metals only overlap slightly and the exchange integral J is small and positive. When the atoms come closer together and the 3d electrons approach one another more closely, the positive exchange interaction, favoring spins, becomes stronger and then decreases to zero. A further decrease in the interatomic distance brings 3d electrons so close together that their spins must become antiparallel, leading to negative J. This condition is called antiferromagnetism. These remarks are particularly relevant to clusters where interatomic distances are modified due to the presence of the surface.

It is commonly admitted today that bulk ground state properties of transition metals are well described by an itinerant electron picture, developed by Stoner and implemented in connection with modern band theory. In contrast, rare earth metals are better described by the localized electron picture. Contrary to the case of 3d wave functions in transition metals, 4f wave functions of rare earths only overlap weakly. The standard model of rare earth magnetism is based upon the approximation that the 4f states are essentially the same in the solid as in free atoms. The electronic configuration is defined by Hund's rules. Since the 4f electrons lie well within the ion cores, the interaction with the environment is adequately represented by local exchange interactions between 4f and (5d, 6s) conduction electrons. As a result, the exchange interaction between the localized 4f spins $\boldsymbol{S}$ is of indirect nature and is mediated by conduction electrons in the RKKY fashion [34,35]. Due to the strong spin orbit coupling in the 4f shell, $\boldsymbol{J}$ rather than $\boldsymbol{S}$ is a constant of the motion. The exchange is then determined by the projection of $\boldsymbol{S}$ on $\boldsymbol{J}$, which is $(g-1)\boldsymbol{J}$ and the exchange energy is proportional to $(g-1)^2 J(J+1)$. Therefore the exchange energy is highest for elements in the middle of the rare earth series. Characteristic exchange energies correspond to temperatures ranging from tens to hundreds of Kelvin. The large magnetic anisotropy of rare earths are due indirectly to unquenched orbital momentum and spin orbit coupling in the incomplete f-shells of atoms. As a result, most properties of rare earths can be explained in terms of a strong competition between the exchange energy and the magnetic anisotropy energy [36].

5.2.2 Implications for Cluster Magnetism

Cluster magnetism will be particularly sensitive to the the local environment of the atoms at the surface. In particular, it is well known that the reduced coordination of surface atoms leads to a band narrowing. Surface sites become more atomic like and increase their local moments. For transition metals, the electronic d band, which is responsible for magnetism can be adjusted to the second moment of the local density of states on atom i. The band width can then be expressed as

$$W_i = W_{\text{bulk}}(Z_i/Z_{\text{bulk}})^{1/2} . \tag{5.1}$$

Assuming equal band width for up and down spins, and assuming that the exchange splitting is the same as in the bulk, the magnetic moment on atom i can be written:

$$\mu_i = (Z_{\text{bulk}}/Z_i)^{1/2}\mu_{\text{bulk}} . \tag{5.2}$$

This relationship is only valid for not too small value of Z_i. It shows that for small clusters, the average magnetic moment is largely dominated by surface atoms since small Z_i implies large μ_i. Atoms in the center of the clusters, have all their near neighbors and therefore $Z_i = Z_{\text{bulk}}$.

On the other hand, the contraction of interatomic distances that are often observed at surfaces may produce a reduction of the total moment. At first glance, these effects appear to balance one another. Since surface relaxations are hard to evaluate in clusters, most calculations only account for the first effect. On the basis of these calculations, speculations arose as to the possibility of magnetic ordering in V_N, Ru_N, Pd_N and Cr_N clusters. Recent measurements have ruled out significant magnetic ordering in Ru_N, and Pd_N clusters [12]. Because of their different sensitivity to local environment, the 3d and 4f magnetism will manifest themselves in quite different ways in clusters. Moreover, the symmetry breaking at the surface may result in large surface-induced magnetic anisotropies. Recent measurements, [9] indicate that rare earth clusters are indeed magnetic and that they exhibit behaviors that are not observed in transition metal clusters.

While magnetic ground state properties of clusters have been calculated with various degrees of accuracy for transition metals (TMs), no first principle calculations have been performed for excited states. In particular, serious problems arise if one wishes to extend the itinerant electron model to finite temperatures and the approach breaks down entirely for the paramagnetic regime above the critical temperature. Therefore at finite temperature, only the localized spin picture is available to explain the magnetic properties. Despite the complexity of the real atomic structure of a ferromagnet, the essential finite temperature properties can be explained quite satisfactorily by this model. In this context, ferromagnetic ordering and magnetic phase transitions can be understood in terms of fluctuations with many length scales [37]. By diminishing the size of a system, one cuts out low lying excitations (long wavelengths). Since spin fluctuations persist at all smaller scales of length, interesting effects are expected when the correlation length of the spin system is no longer commensurate with the linear dimensions of the particle. For magnetically ordered materials, there is a size (typically 20 nm), below which the giant (single domain) magnetic moments of the particles begin to fluctuate in direction at temperatures well below the bulk Curie temperature. The particle looses its spontaneous magnetization in the laboratory reference frame but responds strongly to an applied field. This type of rotation fluctuation of the magnetic moment was first pointed out by Neel [38]. See [39] for a review of this so called superparamagnetic behavior. The switching rate of the magnetization is thermally activated and therefore depends exponentially on the energy barrier U and the temperature T:

$$\Gamma = \Gamma_0 \exp(-U/kT) , \qquad (5.3)$$

where Γ_0 is related to the precession frequency and is estimated to be 10^9 to $10^{13}\,\mathrm{s}^{-1}$. While shape anisotropy accounts for the energy barrier of non-spherical particles, crystal anisotropy is responsible for the variation of energy due to orientation of the magnetization with respect to the crystal axes. When the crystalline contribution to the anisotropy is the dominant term,

$U = KV$, where K is the anisotropy energy (for typical ferromagnets, $K \approx 0.01-1\,\mathrm{J/cm^3}$) and V is the volume of the single domain particle. There is a temperature :

$$T_{\mathrm{bl}} = \frac{KV}{k\ln(\tau_{\mathrm{exp}}\varGamma_0)} \;, \tag{5.4}$$

below which thermal energy fluctuations are too slow to be observable on the time scale of the experiment τ_{exp}. The particles are said to have reached their blocking temperature T_{bl}.

5.3 The Stern-Gerlach Experiments

5.3.1 Experimental Principles

All facilities in use are of the type described by Cox et al. [47]. This technique couples the conventional Stern-Gerlach deflection scheme with laser vaporization cluster source (LVCS) technology and time-of-flight mass spectrometry (TOFMS). The metal clusters are grown in a helium filled source from atoms produced by the pulsed laser vaporization of a sample material. In recent designs, the source retains the clusters in a cavity for 1 to 2 ms (for experimental details and and source operation mode see for example [48]). During this time, the clusters stop growing and thermalize with the source cavity. The helium-cluster mixture then undergoes a free jet expansion into the vacuum, producing a supersonic cluster beam. While the free jet expansion that forms the beam cools the translational and rotational temperatures of the clusters, it has little or no effect on their vibrational temperatures [48]. By restricting the studies to clusters that have reached thermal equilibrium with the source cavity, the clusters vibrational temperature is established to within a few K. The source chamber can be attached to a closed cycle helium refrigerator and electric heater, so that its temperature can be controlled between 60 and 350 K.

In the design of the University of Virginia [48], a mechanical chopper near the source allows only a brief pulse of these equilibrated clusters to pass; it also serves as a reference event for calculating the clusters beam velocity. The beam is collimated and passes through gradient magnet, where the clusters are accelerate perpendicular to their trajectory according to the projection of their magnetic moment onto the field axis. They are deflected in a drift tube before entering a ionization region of a TOFMS. These clusters are ionized by a narrow ultraviolet beam from an ArF excimer laser (193 nm), directed antiparallel to the cluster beam. This laser beam scans back and forth, mapping out the profile of deflected clusters. Once ionized, the clusters are dispersed perpendicular to the beam trajectory and detected in a mass spectrometer. By analizing the spectra of clusters obtained at many positions of the excimer laser beam, it is possible to determine cluster deflections as functions of cluster size, vibrational temperature, and magnetic field. In this way deflection of individual clusters can be measured up to several hundred atoms.

Initial results on transition metal clusters, i.e. Fe and Co [7,8] showed that these clusters deflect exclusively towards increasing magnetic field. They produce a relatively narrow, well defined deflection profile. Therefore the following expression relating the deflection d to the average magnetic moment per atom projected onto the field axis $\langle \mu_z \rangle$ can be used

$$d = L^2 \frac{(1 + 2D/L)}{2mv_x^2} \partial B/\partial z \langle \mu_z(N, B, T) \rangle , \qquad (5.5)$$

where $\partial B/\partial z$ is the magnetic field gradient in the z-direction, L is the length of the gradient field magnet, D is the distance from the exit of the magnet to the TOFMS, m is the atomic mass, v_x is the velocity of the clusters. $\langle \mu_z \rangle$ must be considered as a time average value of the projection of the magnetic moment onto the field axis and depends implicitly on B, T and the number N of atoms in the cluster.

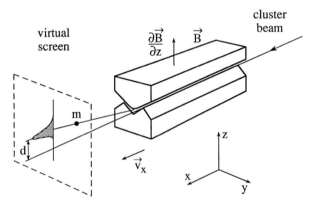

Fig. 5.1. Perspective view of the quadrupole sector pole faces used at the University of Virginia. This configuration offers optimal field gradient homogeneity. Observable parameters are also represented and correspond to (5.5). Schematically shown on the virtual screen is the beam profile after deflection; it contains information on the average magnetic moment but also on the cluster's relaxation

De Heer and coworkers already recognized that the cluster magnetic moments determined in this way were far below the bulk value [7]. The reduced magnetization was contrary to theoretical predictions that lowering the coordination enhances local moments and hence clusters should be more magnetic than bulk. The comparison between theory and experiment seemed quite controversial, specially concerning the temperature dependence of d. It was not until the work of Bucher, Douglass and Bloomfield [8] that the relation between the measured deflection d and $\langle \mu_z \rangle$ became clear. They performed experiments on Co_N clusters and showed that the observed small average deflections could be interpreted by assuming a superparamagnetic relaxation of the clusters in the beam [8]. This interpretation is supported by the theoretical analysis of Khanna and Linderoth [49].

5.3.2 Magnetic Moment Measurements of Unsupported Clusters

Over the temperature range experimentally accessible, the clusters of transition metals behave superparamagnetically. Although each cluster exhibits ferromagnetic order and has a giant magnetic moment, the orientation of that moment fluctuates on the nanosecond time scale. The moment is only weakly attached to the cluster's atomic lattice, and thermal agitation from the vibrational modes causes the moment to explore all possible orientations during the several 100 microseconds the clusters spend in the gradient field magnet.

This fluctuating giant moment responds paramagnetically to an applied magnetic field, yielding a smaller time averaged magnetic moment aligned with that field (in the laboratory reference frame). The applied field modifies the Boltzmann factors weighting the moment's possible orientations, so that the cluster exhibits a magnetic moment per atom $\langle \mu_z \rangle$ that is reduced from its full internal moment per atom μ by the Langevin function $\mathcal{L}$,

$$\langle \mu_z \rangle = \mu \mathcal{L} \left(\frac{N\mu B}{kT} \right) = \mu \left[\coth \frac{N\mu B}{kT} - \frac{kT}{N\mu B} \right], \quad (5.6)$$

where N is the number of atoms in the cluster, B is the magnitude of the applied field, k is Boltzmann's constant and T is the cluster's vibrational temperature. Since the deflection measurement yields a value of $\langle \mu_z \rangle$, the cluster's internal moment per atom μ is obtained by inverting this relationship.

The superparamagnetic behavior because of its clear signature allows a systematic study of internal magnetic moments of clusters. It becomes manifested by the scaling behavior of magnetization as a function of N, B and T. The magnetization per atom of a cobalt clusters increases linearly as a function of magnetic field as shown in Fig. 5.2 for three different cluster sizes and two different temperatures. A linear increase of the magnetization per atom is also observed as a function of clusters sizes. Finally, the magnetization is observed to vary as the inverse of cluster temperature. This scaling behavior of the magnetization as a function of N, B, and T was demonstrated for the first time by Bucher et al. [8] thanks to a new cluster source conceived for equilibrium measurements [48]. A strait line passing through the origin is obtained for a given size N, when $\langle \mu_z \rangle$ is reported as a function of NB/T. The proportionality factor must be the square of a magnetic moment; its value is always in excess of the bulk magnetic moment per atoms. These findings are essential and appear almost like a signature since $\langle \mu_z \rangle / \mu = N\mu B / 3kT$ is the

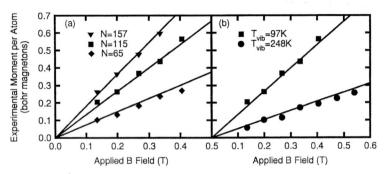

Fig. 5.2. Measured magnetic moment per atom as a function of magnetic field for Co clusters. (a) $T_{\text{vib}} = 97\,\text{K}$ for $N = 6$, 115, and 157. (b) $N = 115$, for $T_{\text{vib}} = 97\,\text{K}$ and 248 K. From [8,10]

first term of the development of a Langevin function of argument $N\mu B/kT$ and hints strongly at the statistical nature of excitations in these clusters. It was Khanna who first emphasized that a superparamagnetic interpretation could fit the experimental data [49]. A magnetic moment per atom of $2.1\,\mu_B$ is obtained for $N = 115$. This value is enhanced over the bulk value of $1.72\,\mu_B$ per atom, in good agreement with the simple prediction outlined in Sect. 5.2. The superparamagnetic regime opened up the possibility of a whole set of experimental studies.

5.3.3 A High Resolution Experiment: Nickel

Measurements with unpreceeding precision have been performed for Ni_N clusters in the size range between $N = 5$ and 100 [40]. These measurement show an oscillatory behavior of the magnetic moment as a function of size (Fig. 5.3). In particular, Ni_5 has the highest magnetic moment per atom while Ni_{13} has a strong minimum. For Ni_5 the measured value of $1.8\,\mu_B$ per atom is close to the value of $1.6\,\mu_B$ per atom calculated by Reuse and Khanna [41] on the basis of density functional (DS) calculation. Reuse and Khanna's predicted that Ni_{13}, which is the first compact cluster (cluster with a complete icosahedral shell), has the lowest moment per atom [42]. They furthermore found the correct trend of decreasing μ between Ni_5-Ni_6 and between Ni_8-Ni_{13}.

For larger clusters other minima occur in the mass spectrum of Fig. 5.3 around Ni_{34} and for Ni_{56}. Indeed, Ni_{56} is close to Ni_{55} that would correspond to a closed shell icosahedron. For clusters bigger than Ni_{13}, one has to compare experimental results with tight binding (TB) calculation. Such TB calculations have been performed up to Ni_{60} by Bouarab et al. [43,44]. Again the general trends are reproduced, in particular the broad minimum centered around Ni_{34} and the close shell minimum at Ni_{55}. The steady increase of $\mu(N)$ above Ni_{28} mainly reflects a variation of the coordination number as analyzed in Sect. 5.2. These authors also studied the effect of variation of the

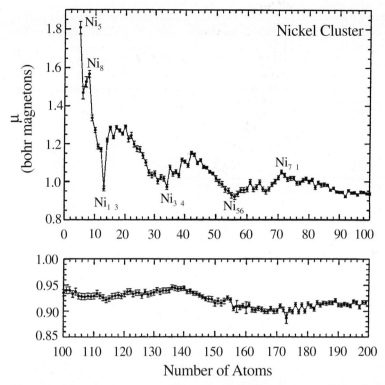

Fig. 5.3. Magnetic moment of Ni_N clusters measured in a Stern-Gerlach experiment by Apsel et al. [40]

interatomic distance. The experimental maxima in $\mu(N)$ are not reproduced however in the TB calculations. The main problem in carrying out the calculations is obtaining the geometry of the cluster. The details of $\mu(N)$ would necessitate a treatment of both the geometry, by minimization of the total energy, and a self consistent calculation of the electronic structure.

5.3.4 Temperature Dependence of the Giant Moments

Up to now we have assumed that $\mu' = N\mu$ is a giant magnetic moment that does not depend on temperature in first approximation (remember that μ is obtained from the measurement of $\langle \mu_z \rangle$ and the knowledge of T, see (5.6)). This approximation is legitimate for small T. Depending on the cluster size $T = 100$ K may already be considered as a low temperature. The true order parameter μ however depends on T. Figure 5.4 shows a typical dependence of $\mu(T)$ calculated from a Heisenberg Monte Carlo simulations [45] for a 55 atom cluster. Individual spins of atoms are completely aligned at zero temperature. The spins in the cluster lose their mutual alignment and the magnitude of the single domain magnetic moment μ' decreases as a function of T. However μ'

does not drop to zero at $T = T_c^{bulk}$, since the spontaneous symmetry breaking at T_c can occur in the thermodynamic limit only (N goes to infinity). The Langevin function still fits perfectly the calculated points $\langle \mu_z \rangle$ as a function of h, in Fig. 5.4b, provided the new $\mu(T)$ is taken from Fig. 5.4a.

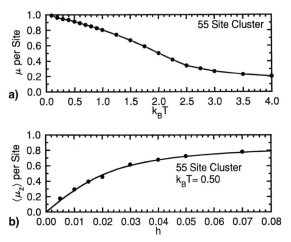

Fig. 5.4. Heisenberg Monte Carlo calculation for a 55-atom cluster. The isotropic Heisenberg model considers an effective coupling $\mathcal{J}$ between spins, which in a molecular field theory, is estimated to be a fraction of the Curie temperature kT_c^{bulk}. Energies are in units of $\mathcal{J}$, the spin-spin coupling energy [45]. (a) Magnitude of the total spin per atom as a function of temperature. (b) Time average projection of the magnetic moment per atom on the field axis at $T = 0.50$ as a function of applied magnetic field. The curve is a parameterless fit by means of the Langevin function

To find the experimental $\mu(T)$, a deflection experiment at low temperature must be done first in order to measure the ground state maximum value of the magnetic moment. Then the measurement can be done for any vibrational temperature. Temperature variation of μ have been studied in Stern-Gerlach experiments by Billas et al. [13] for Ni, Co and Fe clusters. In the case of Ni and Co clusters, and for a given cluster size N, $\mu(T)$ decreases as expected as a function of temperature although not as rapidly as predicted by the Heisenberg model. When the cluster size is increased, the $\mu(T)$ curves more and more match the bulk curve. This is the case for Ni clusters up to a size of about $N = 500$; above this value, $\mu(T)$ does not go to zero, suggesting that significant magnetic ordering subsists to temperatures well above T_c^{bulk} in good agreement with the Heisenberg model. Recent results by Apsel et al. [40] seem to fit better the model for low cluster temperatures. On the other hand, the behavior of Fe$_N$ clusters is rather atypical (Fig. 5.5c). The temperature at which a significant drop of $\mu(T)$ is observed is well below $T = T_c^{bulk}$, and surprisingly the value of $\mu(T)$ increases again for higher $T \geq T_c$. This

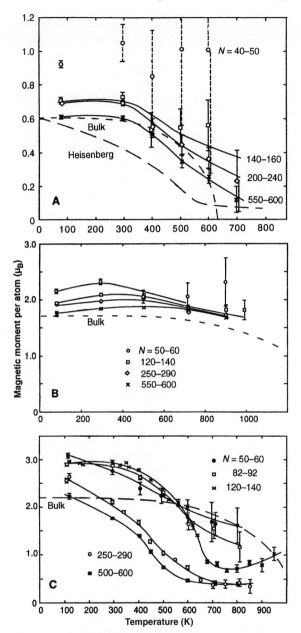

Fig. 5.5. Temperature variation of μ as studied in Stern-Gerlach experiments by Billas et al. [13] for Ni, Co and Fe clusters. *Solid line* are guides for the eye. Nickel clusters converge to the bulk properties for $Ni_{500-600}$ at low temperatures. A slight initial increase of μ is observed in Co clusters. The anomalous behavior of the magnetic moment of Fe clusters is attributed by the authors to a crystal phase transition where the transition temperature decreases with increasing cluster size

effect was suggested to be due to a structural transition interfering with a purely magnetic transition [13]. Within spin fluctuations theory, Pastor et al. [46] could show that short range magnetic ordering within the clusters is responsible for the non zero magnetic moment observed experimentally above $T = T_c^{\text{bulk}}$ for Ni and Co.

Another confirmation of the critical behavior in finite systems is found in Gd_N clusters [9,10]. Due to the relatively low critical temperature of bulk Gd, "high temperature" effects can be studied at room temperature in Gd_N clusters. The fact that these clusters remain magnetic (they deflect) at temperatures well above the Curie temperature of the bulk has direct implication on the temperature dependence of $\mu(T)$. Although detailed measurements of $\mu(T)$ have not been performed yet for rare earth clusters, the observation that $\mu(T)$ still has a finite value at temperatures of $2T_c^{\text{bulk}}$ again confirms the prediction of the simple model depicted in Fig. 5.4.

5.3.5 Clusters of Non-Ferromagnetic Transition Metals

One of the fascinating results in the field of magnetism of small clusters is the discovery that magnetic entities can be built out of materials that are not magnetic in the bulk state. An illustration of such an effect is the case of Rh clusters of a sizes below 60 atoms [12]. Magnetic moments per atoms follow an oscillatory behavior as a function of size, and moments as high as $0.9\,\mu_B$ per atoms are measured for Rh_{10}. Maxima also occur for Rh_{15} and Rh_{19}. Contrary to the case of Co, Ni, and Fe the magnetic moments of Rh clusters only vary over a small size range, for $N > 20$ the magnetic moment has dropped already below $0.2\,\mu_B$. An early theoretical prediction that Rh clusters could be magnetic was made by Reddy et al. [50] on the basis of LSDA. Later, calculations have been done on a wider size range by many authors [51] and qualitative agreement was obtained with the experiment. Most calculations were able to reproduce the peaks; however, scattering of the magnetic moment values between different calculations remains rather important. One of the main difficulties in theoretical studies is their strong sensitivity on the geometrical arrangement of atoms and on nearest neighbor distances both of which are generally not known *a priori* [51].

There is a number of theoretical predictions which indicate that other clusters of non-ferromagnetic TMs may posses significant permanent magnetic moment. Reasonable candidates are Pd_N, Ru_N, V_N and Cr_N clusters. However, experiments along the same line on these elements failed to detect any measurable deflection [12,52]. This is not to say that these clusters are not magnetic at all, but upper limits can be set by the experiment. For the case of palladium, $0.4\,\mu_B$ per atom would still be detectable for Pd_{13} while $0.13\,\mu_B$ per atom could be detected on Pd_{105}.

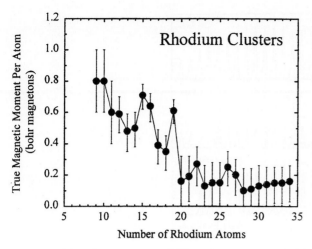

Fig. 5.6. Magnetic moment of Rh_N clusters measured in a Stern-Gerlach experiment by Cox et al. [12]

5.3.6 Locked Moment Clusters and Spin Canting

Clusters of gadolinium and terbium have been found to exhibit either superparamagnetism, as discussed earlier for transition metals, or "locked-moment" behavior, depending on the cluster size and vibrational temperature [9,10,55]. Locked moment behavior occurs when sufficient crystal anisotropy exists in a cluster to constrain the orientation of the spontaneous magnetization vector along an "easy axis". The relevant criterion that differentiates between clusters that are superparamagnetic and those that have locked magnetic moment is the ratio between thermal vibration energy kT_{vib} and magnetic anisotropy energy KV. The anisotropy provides an energy barrier that resists the reorientation of the magnetization vector. In some clusters this energy is small and is overcome easily, even at the lowest temperature accessible experimentally. In other clusters, the anisotropy energy is large enough that it is not overcome even at room temperature or above.

Clusters that have large anisotropies, behave as tiny permanent magnets. Because their magnetization vector is obliged to move with their cluster lattice, they tumble and rotate as a single unit without thermal relaxation. When placed in an external magnetic filed, a spinning cluster with a locked magnetic moment will precess and nutate much as a macroscopic magnetized spherical rotor would do when placed in a magnetic field.

Gd_N clusters exhibit both, superparamagnetism and locked-moment behavior. Many Gd_N clusters have sufficient coercivity to block reorientation of the magnetization vector even at room temperature. There are several notable exceptions: Gd_{22}, Gd_{30}, and Gd_{33} are predominantly superparamagnetic even at the lowest temperatures studied ($T = 95\,\text{K}$) [9]. The size dependence of the magnetic behaviors of gadolinium clusters have lead to

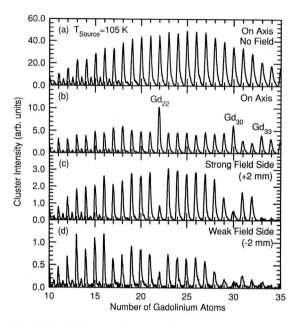

Fig. 5.7. Evidence of magnetic magic numbers in the mass spectra of Gd_N clusters [9]. (a) Mass spectrum of clusters on the undeflected beam axis without any magnetic field. (b), (c), and (d) are spectra of clusters when a field of 0.132 T and a gradient of 50 T/M are applied. (b) shows the clusters that remain on the beam axis, while (c) and (d) show the clusters that are deflected 2 mm to strong field and 2 mm to weak field respectively

the appearance of striking *magic numbers* in the spectra of clusters observed following a magnetic filter (Fig. 5.7). Clusters that are superparamagnetic are deflected quite differently from those that have locked magnetic moments. While the superparamagnetic clusters deflect uniformly towards high field, leaving their initial deflection profile almost unchanged (what is meant by deflection profile is shown schematically in Fig. 5.1), the locked moment clusters are instead characterized by an extreme spatial spreading. An example of such a behavior is Gd_{21} (Fig. 5.7). In fact, both behaviors may also be observed simultaneously for clusters of the same size. For example, at 95 K many Gd_{22} clusters start to exhibit a locked moment character. The cluster beam then contains a mixture of two separate, distinguishable types of Gd_{22} clusters. It is not clear what distinguishes the two groups of otherwise identical clusters. One reason may be the presence of structural isomers.

Superparamagnetic Gd_N clusters have internal magnetic moments per atom that are substantially less than the bulk value of $7.63\,\mu_B$. In the case of Gd_{22}, moments comprised between 3.0 and $3.9\,\mu_B$ per atom are measured in the temperature range from 95 K to 250 K. It was suggested [48] that this reduced moment may be due to fact that the individual magnetic moments are not necessarily aligned, but may be at an angle with the easy axis of magnetization. This effect is actually well known from bulk rare earths [53] and may be accentuated in clusters, where the single ion behavior is very sensitive to the surroundings. This assumption was confirmed later, on the basis of first principle calculations on a Gd_{13} which led to a spin canted configuration [54].

While some cluster sizes seem to prefer superparamagnetism, most gadolinium clusters exhibit predominantly locked-moment behavior at low vibrational temperatures, down to 95 K. As the vibrational temperature increases, some of these clusters become entirely superparamagnetic, after going through a temperature range at which an intermediate behavior is observed. This observation is consistent with the idea of the overcoming an activation energy for the reorientation of the magnetization vector. The locking-delocking mechanism is also most probably what was observed in the marginally superparamagnetic behavior observed on supersonically cooled Fe_N clusters [11].

The magnetic behavior of terbium Tb_N clusters is very similar to that of Gd_N clusters [55]. Terbium like most odd numbered rare earths, is monoisotopic. This feature simplifies the mass spectrum of terbium and makes it possible to distinguish pure terbium clusters from those that include light impurity atoms. Thus, the effect of a single oxygen atom on the magnetic behavior of the clusters can be studied. With the exception of Tb_{22}, the oxygen atom has no measurable effect on the magnetic behavior. For Tb_{22}, which is superparamagnetic, adding a single oxygen atom causes the magnetization vector to lock to the lattice. $Tb_{22}O$ does not become superparamagnetic until its vibrational temperature exceeds 250 K.

5.4 Interpretation of the Beam Experiments

When a cluster enters the inhomogeneous field in a Stern-Gerlach experiment, its magnetic moment starts to exchange angular momentum with its lattice in order to satisfy the conservation of the component of the total angular momentum about the field axis L_z. The fact that the angular momentum $\boldsymbol{L}$ of a cluster is related to its magnetization $\boldsymbol{M}$, only reflects a general symmetry constraint since the angular velocity and the magnetization are both axial vectors whose signs vary in the same way upon time inversion. A macroscopic illustration of this effect is demonstrated in a vivid experiment proposed by Einstein and de Haas [56], where a piece of metal suspended in a magnetic field starts to rotate upon field inversion. The gyromagnetic part of the angular momentum acquired upon magnetization is given by

$$(L_{gm})_k = \sum_l \beta_{kl} M_l \,, \tag{5.7}$$

where $k, l = x, y, z$. The magnetomechanical ratio tensor is then defined by $g_{kl} = \beta_{kl}^{-1}$. In this connection, we should mention that the Einstein-de

Haas experiment and its dual, the Barnett experiment, have been used to measure g values of various metals and alloys [57]. Both of these experiments demonstrate the spin nature of ferromagnetism.

The continuous adjustment of the cluster's angular velocity to the new conditions takes place during the whole flight time of the cluster through the magnet. In practice, as was mentioned already, two extreme situations can be distinguished. In the case of transition metal clusters, fluctuations play the dominant role and alignment of the magnetic moment with the magnetic field occurs by vibrational relaxation (entirely internal to the cluster). The projection of the magnetic moment onto the field axis is given by the laws of statistical mechanics [8,49]. For the majority of rare earth clusters, with $N \leq 100$ and for not too high temperatures ($T_{\text{vib}} \leq 300$ K), the magnetic moment $\boldsymbol{M}$ is tied to an easy axis of the lattice. The behavior is then essentially dictated by the relative importance of the magnetic energy and the rotational energy of the clusters in the magnetic field [9]. The clusters are subject to a complicated precession and nutation behavior and final deflection essentially depends on their state $(\boldsymbol{L_0}, \boldsymbol{M_0})$ when they initially enter the magnet. Experimentally, any kind of mechanism that tends to lock the magnetic moment to the lattice manifests itself by a dramatic broadening of the deflection profile.

5.4.1 Magnetic Anisotropy

The magnetic anisotropy defines the low temperature orientation of the magnetic moment $\mu' = N\mu$ of the cluster with respect to its lattice. This property is of considerable technological importance for magnetic recording. It is generally believed that each particle can store one bit of information. To be attractive for storage applications, this resulting anisotropy must be sufficient to lock moments along a preferred direction at room temperature. The magnetic anisotropy energy can be written phenomenologically as

$$E_{\text{an}} = V[K_2 \sin^2 \theta + K_4 \sin^4 \theta] , \qquad (5.8)$$

where V is the volume, θ is the angle between the easy axis and the magnetization, K_2 and K_4 are the first and second order anisotropy constants. This expression is valid for uniaxial anisotropy and neglects higher order terms. The spin-orbit coupling is responsible for the magnetocrystalline anisotropy contribution to this expression. The equilibrium direction of magnetization is obtained by energy minimization with respect to θ. Accurate computation of relativistic effects [58] giving rise to anisotropies are notably difficult and have just begun for small clusters [59].

5.4.2 Coupling Between Magnetic Moment and Lattice: Deflection Profile

In clusters that behave superparamagnetically, each cluster explores the Boltzmann distribution of possible magnetization vector orientation. Whenever the time taken to measure the average projection of the magnetization vector on a given axis is long compared to the time needed to explore the Boltzmann distribution, individual clusters will appear identical to one another. As an ensemble of superparamagnetic particles traverses a gradient magnet, every particle exhibits the same average magnetization and every cluster experiences the same deflection giving rise to a narrow deflection profile.

But clusters that have their magnetization vectors tied to their lattices are distinguishable and remain distinguishable throughout the experiment. They are distinguished by their initial orientations and angular momenta. Unlike superparamagnetic particles, these locked moment particles cannot exchange angular momentum between their spin systems and their ion lattices. Each particle moves and rotates as a unit, without any internal relaxation. The classical dynamics of motion of such a cluster depends on its initial angular momentum vector and on the relative orientation of that vector with respect to its magnetization vector. As soon as this particle is placed in a magnetic field it feels a torque and begins to precess and nutate. This inequivalence of nominally identical clusters leads to broad deflection profiles (defined schematically in Fig. 5.1). A clusters whose dynamics causes its magnetization vector to remain pointed almost parallel to the magnetic field will deflect towards strong field while a cluster that is most often antiparallel will deflect toward weak field. An ensemble of clusters will experience a wide range of deflections, depending on the rotational temperature of the clusters and their internal magnetic moments. This inhomogeneous broadening of the deflection profile is the principal signature of the locked-moment behavior.

In the case of locked moment (strong anisotropy), the deflection profiles contain information not only on the clusters magnetic moments but also on their rotational temperatures $T_{\rm rot}$. One important parameter contributing to the shape of a deflection profile is the ratio of the magnetic energy over the rotational energy $\mu B/kT_{\rm rot}$. When the magnetic and rotational energy are close to one another, the dynamics of motions are very complicated and must be followed numerically. Behavior of a statistical ensemble of locked moment clusters has been studied in order to determine their distributions of time averaged magnetic moment projection on the magnetic field axis [10]. When $\mu B/kT_{\rm rot} \ll 1$, rotational energy dominates the dynamics and the distribution of time-averaged magnetic moment is symmetric about zero. When $\mu B/kT_{\rm rot} \gg 1$, magnetic energy dominates and the distribution is heavily skewed toward alignment with the applied field.

The shape of the deflection profiles obtained in this manner depend only on the balance between magnetic and rotational energy $\mu B/kT_{\rm rot}$. The scale of the deflection depends on a single factor determined by internal magnetic moment per atom, the magnetic field gradient, and time during which the clusters are permitted to deflect. With this approach it is possible to fit experimental deflection profiles (see for example the case of Gd_N clusters) with a quite good agreement [9]. A rotational temperature of 6 K is obtained for Gd_{21}. At vibrational temperatures near the transition between locked-moment and superparamagnetic behavior, clusters produce deflection profiles that are intermediate between the broad distribution of locked-moment model and the narrow distributions of the superparamagnetic model. Near the transition the magnetization vector is exploring substantial portions of the crystal anisotropy energy surface and is neither superparamagnetic nor locked. A calculation using finite anisotropy barrier has been done by Jensen et al. [60]. By solving the Bloch equations, $\langle \mu_z \rangle$ has been calculated as a function of the anisotropy field and kT_{rot}. See also [61]. They seem to explain to some extent the atypical results for Fe_N clusters [7] where a significant rotational contribution is involved.

5.5 Growth Kinetics of Clusters on Surfaces

Adsorption of atoms on a surface from the gas phase is a non-equilibrium process. The system made of a two dimensional adatom lattice gaz is temporarily supersaturated and tries to restore equilibrium by condensing into islands. As a result, growth can be viewed as a non-equilibrium phenomenon governed by competition between kinetics and thermodynamics. An atomistic view of the processes involved in adatom diffusion and attachment can be found in [22,23]. In this section, we summarize the ways to manipulate the growth kinetics in order to tune the density, size and shape of clusters. This section does not imply any reference to self-organization. For sake of simplicity, we follow the fate of metal atoms adsorbed on perfect, single-crystal, metal surfaces prepared under the best condition of ultrahigh vacuum (UHV). Furthermore, we do not go into the details of atomic exchange leading to unwanted alloy formation.

5.5.1 Tuning the Clusters Density

The basics of nucleation and growth on surfaces are presented on the example of silver adatoms on Pt(111). Although it is not magnetic, this system has been studied more extensively than any other [62,63]. Figure 5.8 shows variable temperature STM images of Ag islands on Pt(111) grown and imaged at various substrate temperatures (between 80 K to 110 K); the silver coverage is 0.12 ML. The continuous diagonal lines in the images are monatomic steps of the Pt(111) substrate and the bright zones are the monatomic thick silver structures. As can be seen, the density of islands drops markedly when the temperature is increased while the average size of the islands increases. There is even a temperature (170 K in this example) at which no islands are nucleated anymore on the terraces and silver condenses at the steps of the platinum substrate (step flow). Silver on Pt(111) is representative of islands formed by two-dimensional isotropic diffusion.

A possible scenario of islands formation and growth is as follows. When an incoming adatom is adsorbed on the surface from the gas phase it may have

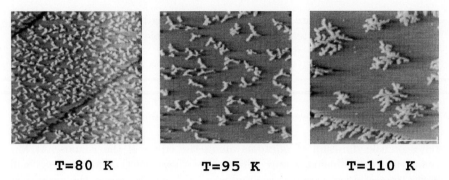

Fig. 5.8. STM topographs of low coverage Ag submonolayer structures ($\Theta = 0.12$ ML) on a Pt(111) surface, grown and imaged at the temperature indicated. The lateral scale is given by the white bar, representing a length of 20 nm [62]

two distinct destinies: (i) it may meet another adatom during its thermally activated random walk and form a stable nucleus (an immobile dimer), or (ii) it may be directly incorporated into an already existing island. The later operation does not contribute to an increase in the number of islands on the surface; it just increases the size of an existing island. The first event is much more frequent in the first stage of the growth process while the incorporation into an already existing island will dominate at longer time. At very long time, no new islands are formed and an incoming adatom diffuses to an existing island before it has the opportunity to meet another adatom. As a result, the island density saturates and we can define a capture area for each island. The average separation between islands then provides a good measure of the adatom diffusion length Λ_a. Nucleation at steps occurs when the average diffusion length before encounter of an island becomes bigger than the extension of one terrace. An analytical expression of this saturation density of islands N_{sat} has been proposed by Venables [64]:

$$N_{\text{sat}} = \eta(\theta)\left(\frac{R}{\nu_0}\right)^{\frac{i}{(i+2)}} \exp\left[\frac{E_i + iE_d}{k_\text{B}T(i+2)}\right], \quad (5.9)$$

where η is a slow varying function of the coverage Θ, R is the deposition rate and ν_0 is the attempt frequency of the order of 10^{13} to 10^{14} Hz. The size of the critical nucleus is i and E_i is its binding energy. $(i+1)$ is the smallest island still stable at a given temperature. When the critical nucleus is one, E_i becomes zero and we get the exponent $E_d/3$. The formula reflects the fact that the adatom diffusion is thermally activated with an energy barrier E_d. In this limit, the adatom migration barrier can be directly obtained by exploiting the results of variable temperature STM. When the saturation density of islands is reported as a function of $1/T$ in an Arrhenius plot, the slope of the straight line directly provides E_d. It is of advantage to work at low temperature, in the regime where the critical nucleus is one, because then we do not have to worry about the binding energy of dimers, trimers, etc.

The variable temperature STM approach is among the most precise method to determine E_d. An extensive review of this technique together with a table of diffusion barriers are given in [65].

5.5.2 Island Shapes

Up to now we have shown that the density of islands relies on the diffusion of adatoms on the bare surface. We show now that the shape of these islands relies on the diffusion of adatoms along the edges of the islands. At low temperatures, diffusing atoms stick where they hit; therefore islands will show fractal structures because diffusion is hindered along the perimeter of the island (see Fig. 5.8c). The fractal dimension of the islands is 1.78 which is very close to what we get from the computer simulation within the diffusion limited aggregation model (DLA) [66]. This ideal situation of frozen perimeter, is realized in practice only at sufficiently low substrate temperatures. The incidence of substrate temperature and flux of the incoming adatoms, on the final shape has been analyzed in detail recently [65] and a transition from randomly ramified to dendritic islands has been evidenced. At higher temperatures at which diffusion of adatoms is activated around the perimeter [67], rearrangements take place and we may obtain compact structures which are close to a thermodynamic equilibrium.

For example hexagonal aggregates have been obtained by depositing Pt atoms on Pt(111) at 450 K and it was shown [68] (Fig. 5.9) that there is a transition from hexagonal to triangular islands as a function of growth temperature. The results were explained on the basis of a well known crystallographic fact, namely that a compact island on a (111)-type substrate is limited by both, edges with a 111 face and edges with a 100 face. The triangular shapes then appear as a result of preferential diffusion along one type of edge and accumulation at the corners. Triangular cobalt islands are observed also when deposited at 300 K on Cu(111). In this case however fcc

 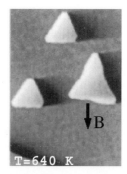

Fig. 5.9. Island shapes on Pt(111) after deposition at substrate temperatures of 455 K and 640 K. STM images 77 nm × 110 nm and 230 nm × 330 nm respectively and $\Theta = 0.15$ ML [68]

and hcp stackings are nearly degenerate in energy and two orientations of the triangles, rotated by 60 degrees, coexist [69].

Under certain circumstances (anisotropic diffusion or anisotropic bonding) [70–72] atomic chains and stripes can be grown. In this case, particular symmetries of the surface may play a leading role. For example by favoring the easy diffusion of adatoms along potential wells or grooves. The anisotropic diffusion can be exploited to tailor highly elongated metastable islands. The simplest example of an anisotropic substrate is a (110) surface of a fcc-crystal (Fig. 5.10). Such a substrate is made of compact atomic rows, along the $[1\bar{1}0]$ direction, separated by channels. In a favorable case, the adatoms will diffuse preferentially within the channels in a linear type of random motion, at least at not too high temperatures. The linear chains form spontaneously by aggregation of adatoms diffusing along the channels.

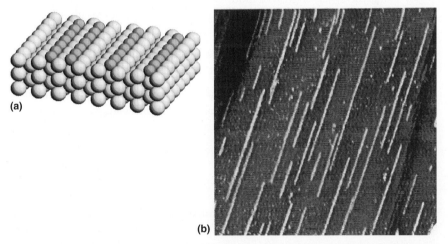

Fig. 5.10. (a) Schematic view of linear atomic chains formed in the grooves of the (110) face of a fcc crystal. (b) STM image with 0.1 ML of copper deposited at 300 K on Pd(110). Images 120 nm × 120 nm [71]

An example of this simple picture has been evidenced in a variable temperature experiment of Cu (0.1 ML) adsorbed on Pd(110) in the temperature range from 265 K to 350 K. At temperatures below 300 K, linear chains several 10 nm long, "monatomic" in width, grow spontaneously along the $[1\bar{1}0]$ direction. Above 300 K, 2D islands that are still elongated along the $[1\bar{1}0]$ direction start to form. Simultaneously, the surface density of islands drops markedly as a function of temperature. It was shown from LEED experiments that these Cu islands grow pseudomorphically on Pd(110). Several examples of growth mechanisms of linear chains have been described in the literature [73,74].

5.6 Thermodynamic Growth Modes

Up to now we only addressed the question of the shapes of the first monolayer of adislands. The reasons why clusters grow in the third dimension and particular shapes are favored as a consequence of strain relaxation is summarized in what follows. In particular, to synthesize well defined metallic nanostructures (organized or not), on a surface, we will see that the formation of flat and defect free layers must be avoided.

5.6.1 Growth Criteria

One usually distinguishes three main growth modes of nanostructures of adsorbed atoms on surfaces [75] (see Fig. 5.11).

(i) The layer-by-layer growth mode (so-called Frank-Van der Merwe mode), in which a layer n grows atomically flat and is completed before the next layer $n+1$ starts growing.

(ii) The Stranski-Krastanov mode: in a first stage the growth occurs layer-by-layer for one or several layers and is then followed by a 3D growth.

(iii) The 3-dimensional (3D) growth mode (also called Volmer-Weber mode) in which crystallites grow vertically, keeping a reduced contact area with the substrate, rather than expanding laterally.

In a simple theory of growth [75] one considers a systems at thermodynamic equilibrium. This theory neglects all kinetic effects occurring during the nucleation processes mentioned above and does not account for alloying. Two equilibrium situations can be considered: in the first one, the atoms A completely cover the surface of the substrate S while in the second one, they form a 3D crystallite of material A (bulk-like), leaving free the major part of the substrate surface S. The energy difference per surface area of the two situations is given by:

$$\Delta\gamma = \gamma_A + \gamma_{AS} - \gamma_S , \qquad (5.10)$$

where γ_A, γ_S, γ_{AS} are respectively the surface energies of the adsorbate, of the substrate and the interfacial energy.

If $\Delta\gamma < 0$, energy is gained when layers cover the surface, hence a layer-by layer (Frank-Van der Merwe) growth mode will occur.

If $\Delta\gamma > 0$, the growth will be three-dimensional (Volmer-Weber).

Here we neglected the surface tensions of the different facets that bound the crystallites (the value of the surface tension depends on the facet orientation) which may also determine the equilibrium shape of the adsorbate (Wulff theorem) [76–79]. Mezay and Giber calculated the surface energies for polycrystalline metals, using experimental data [80]. More recently, metal surface energies have been computed with *ab initio* methods, taking into

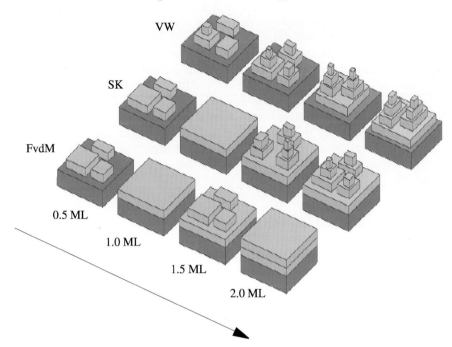

Fig. 5.11. Schematic view of Frank-Van der Merve, Stranski-Krastanov and Volmer-Weber growth modes as a function of coverage

account the crystallographic orientation of the surface [81]. The interfacial term γ_{AS} is difficult to estimate and was often neglected in the past [82]. For metal on metal growth, it is *a priori* not possible to neglect the interfacial energy, particularly when $\gamma_A \approx \gamma_S$. The sign of $\Delta\gamma$ is then mostly determined by γ_{AS}. To obtain flat interfaces, A has to wet S but S must also wet A. This is theoretically only possible when γ_{AS} is non-zero. A microscopic approach of wetting, based on tight binding calculations, has been developed by Gautier and Stoeffler [83]. Their results mainly confirm the phenomenological approach of Bauer and van der Merwe.

5.6.2 Elastic and Structural Considerations

In thermodynamic models, usually only pseudomorphic layers are considered, i.e. the adsorbate is supposed to adopt the substrate structure. In order to improve this model, an elastic contribution arising from the misfit energy between the substrate and the adsorbate can then be introduced, as proposed by Kern et al. [78]. The elastic misfit energy was added explicitly and the spreading energy $\Delta\gamma(z)$ was assumed to decay exponentially towards $\Delta\gamma$. Recently, Muller et al. allowed the successive layers to relax [84]. They showed that the strain drives the transition from the layer-by-layer growth towards the Stranski-Krastanov mode, in agreement with earlier molecular dynamic simulations using Lennard Jones potentials [85]. After some critical number of 2D pseudomorphic layers have grown, the accumulated strain has to be relaxed. As a result, the islands grow vertically instead of expanding laterally and dislocations appear. The dislocations modify the shape of the growing 3D crystallites. The larger the elastic energy, the larger will be the shape ratio $\frac{\text{height}}{\text{lateral-size}}$ of the islands. The substrate is also affected by the strain release since it is dragged by the relaxing island [86]. Therefore, in order to grow well defined entities on surfaces (aggregates, clusters), one should associate elements having a large misfit and a large substrate to adsorbate stiffness ratio [79].

5.7 Organized Growth

Two types of self-organized islands can be distinguished. The first type proceeds through a cooperative growth where the atoms of two adjacent islands interact at a distance and rearrange during the growth process; it can be termed as evolutionary, dynamic self-organization (very often occurring in the Stranski-Krastanov growth mode at thermal equilibrium; see Sect. 5.6.1). This lateral exchange of information (mesoscopic correlation of elastic origin) between atoms allows to grow hetero-structures that self-organize in the third dimension. A nice example of such growth can be found in semiconductors [87]. In the second type of growth the substrate behaves as if it was inert. Surface reconstructions and vicinal surfaces are used as templates. These two processes are not mutually exclusive, although in the case of metal structures of interest to us, the second type is more frequent.

Surfaces may reconstruct spontaneously as a result of free energy minimization. This means that with respect to their positions in the bulk crystal, surface atoms adopt new equilibrium positions. An example of such a reconstruction is the chevron reconstruction of the Au(111) surface. It produces rearrangements of surface atoms at the nanoscopic scale that may appear as a modulation of the topmost atomic layer (see Sect. 5.7.2). However dislocation networks that are interesting for self-organized growth are most easily induced by hetero-epitaxy where the lattice mismatch between two different materials is exploited.

5.7.1 Incommensurate Modulated Layers

Modulated phases can be studied within a simple one-dimensional model proposed in the thirties by Frenkel and Kontorova [88]. This model takes into account the competing interactions between a substrate potential and lateral adatom interactions. A chain of atoms coupled by harmonic springs is placed in a cosine substrate potential of amplitude V and periodicity a. The equilibrium separation of atoms in a chain is b and the force constant of the springs is K. The energy of the system is then given by:

$$H = \sum_n \frac{K}{2}(x_{n+1} - x_n - b)^2 + \sum_n V[1 - \cos(2\pi\frac{x_n}{a})], \quad (5.11)$$

where x_n is the position of the n-th atom. Frank and van der Merve (FvdM) solved this equation analytically within a continuum approximation [89]. They replaced the index n by a continuous variable and x_n by a continuous function $\varphi(n) = (2\pi x/a) - 2\pi n$. The problem contains three parameters: the misfit $\delta = (b - a)/a$ and the two constants K and V. The results show that for slightly differing lattice parameters of chain and substrate potentials (small δ), the lowest energy state is obtained for a system which consists of large commensurate domains, separated by regularly spaced regions of bad fit (Fig. 5.12). The regions of bad fit are called misfit dislocations, solitons or domain walls. They can be considered as collective long period lattice distortion waves, which are excitations of the commensurate ground state. In the continuum limit the ground state satisfies the time independent sine Gordon equation:

$$\frac{d^2\varphi}{dn^2} = pA\sin(p\varphi), \quad (5.12)$$

with $\sqrt{A} = \frac{2\pi}{b}\sqrt{V/K}$, and p the commensurability. One solution of this equation is the solitary lattice distorsion:

$$\varphi(n) = \frac{4}{p}\arctan[\exp(pn\sqrt{A})], \quad (5.13)$$

the so-called soliton. This solution, shown in Fig. 5.12, describes the domain wall at $n = 0$ between two adjacent commensurate regions.

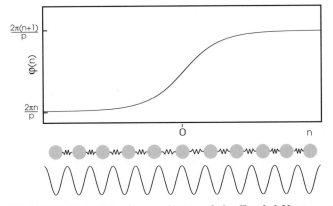

Fig. 5.12. Uniaxial soliton solution of the Frenkel-Kontorova model. It describes the domain wall located at $n = 0$ separating two adjacent commensurate regions. In this model, the width of the domain wall is $L = 1/(p\sqrt{A})$

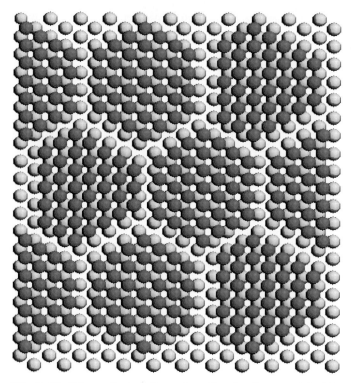

Fig. 5.13. Schematic two-dimensional hard sphere model of fcc and hcp commensurate domains separated by domain walls

For small $\sin\varphi$, Eq. (5.12) can be linearized yielding an exponential decay and therefore an exponential repulsive interaction between walls. In two-dimensional systems, walls are lines with a finite width. Since there are three equivalent directions in a compact crystallographic plane, the domain walls can cross. In the case of a (111)-fcc crystal, adjacent domains may correspond to different stackings of the topmost layer which is either fcc of hcp (see Fig. 5.13). The average period of the dislocation pattern is then given by $D = b/(b-a)$, as a function of the lattice constants b of the film, and a of the substrate.

5.7.2 Atomic Scale Template

Self-organization of metal clusters and islands on surfaces strongly relies on the occurrence of surface reconstruction and strain relaxation patterns. Dislocation networks similar to those shown schematically in Fig. 5.13 have been obtained on hetero- and homo-epitaxial systems. Figure 5.14 shows the dislocation pattern formed by Ag bilayers on Pt(111). Another, widely studied example, is the surface reconstruction of Au(111) [90,91]. Visible in Fig. 5.15

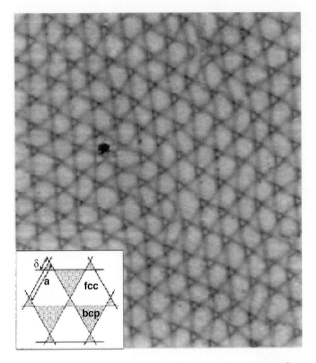

Fig. 5.14. STM image of the domain wall network obtained after annealing the Ag-bilayer on Pt(111). The inset shows a model for the trigonal domains [63]

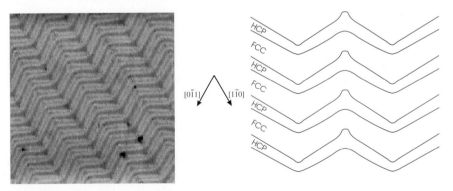

Fig. 5.15. Zigzag reconstruction of the Au(111) surface. *Left*: STM image 100 nm × 100 nm. *Right*: schematic view of the domains

are the bright contrasts of the zigzag domain walls that produce corrugations in the STM images. These discommensuration walls are limiting fcc and hcp stacking domains of the topmost atomic layer which form spontaneously by strain relaxation. The registry of surface gold atoms varies from hollow sites of the fcc to the hollow sites of the hcp stacking (the fcc regions are wider than the hcp regions). The fcc to hcp transition (the discommensuration wall) appears as ridges in the STM images since surface atoms near bridge sites rest about 0.2 Å higher than in hollow sites.

Due to its particular conformation, the zigzag reconstruction of Au(111) leads to singularities that can best be illustrated as single atomic sites dislocation (five nearest neighbors instead of six) located at the elbows of the chevron reconstruction [16]. The dislocations are distributed on a rectangular lattice with a unit cell 7.5 nm × 14 nm. Although this reconstruction has first been evidenced experimentally, it is fairly well understood today thanks for example to molecular dynamics simulations [91].

Among other potentially interesting systems with stress relaxation patterns, lets mention Au on Ni(111) [92], Cr on Pt(111) [93] and Ag on Cu(111) [94], which have been investigated both from the experimental and theoretical point of view.

5.7.3 Self Organization

As far as self-organization of metal clusters is concerned, two approaches have been exploited to date: nucleation on ordered point dislocations and nucleation by capture within cells limited by dislocation lines. Some attempts to use these templates for the self-organization of magnetic systems have been made but to our knowledge, magnetic measurements could be performed to date only on Co/Au(111). Therefore, we focus on Co/Au(111) which has been studied in detail.

Nucleation on Ordered Dislocations. It has been observed that the Au(111) surface may be used as a template to grow organized metal clusters of Ni [16], Co [17], and Fe [95]. The point dislocations at the elbows of the chevron reconstruction (see Fig. 5.15) act as nucleation sites for adatoms adsorbed from the gas phase. On the example of Ni, it has been suggested that, at its initial stage, this mechanism involves a site exchange between one Ni and one Au atom [96]. This substituted atom will then act as a nucleation site for further incoming Ni atoms. In the same way, self-organized cobalt bilayer clusters containing 300 atoms each, can be synthesized (see Fig. 5.16). If one would be able to store information in these dots, one would reach a storage density of 10^{12} bits · cm^{-2}, namely 10^3 times higher than the highest storage densities reached today.

These clusters are stable up to 400 K in UHV environments. Above this temperature, the clusters burrow into the gold substrate, by simultaneously expelling gold atoms. This is due to the low surface energy of gold which tends to encapsulate the cobalt clusters [26]. Annealing below 600 K, does not perturb significantly the cobalt clusters since the magnetic properties are

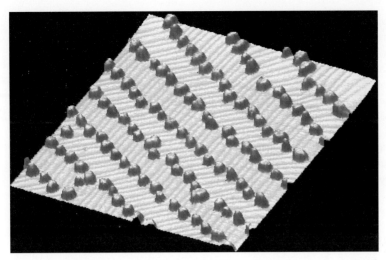

Fig. 5.16. Cobalt clusters made of 2 Co-monolayers, approximately 300 atoms each, self-assembled on the zigzag reconstruction of the Au(111) surface. Image 100 nm × 100 nm. From [107]

preserved. For *ex situ* measurements and potential applications, Co clusters and films can easily be stabilized by a protective film of a few gold monolayers [26,97].

Confined Nucleation. Confined nucleation was demonstrated recently [18], on the example of submonolayer Ag structures grown on strain relaxation patterns of Ag/Pt(111) (see Fig. 5.14). Since the dislocation lines are repulsive for the diffusing silver atoms, confinement of silver atoms inside the unit cell in which they have landed (capture area) takes place. Due to the fact that the Ag adatoms are sufficiently mobile they will form one, and only one, island per unit cell located at the center of the cell. Long range repulsion at dislocations together with preferred binding to fcc areas creates in each unit cell a local adsorption minimum to which the atoms are guided. The island size distribution for this type of nucleation is binomial and, therefore, significantly sharper than for nucleation on isotropic substrates (see Sect. 5.5). The feasibility of this approach has been demonstrated also for Fe on Cu/Pt(111) [18]. It should be mentioned however, that this route of synthesizing nanostructures has been tested at low temperature only (below 300 K).

5.7.4 Periodic Patterning by Stress Relaxation

When growing adatom islands on a surface, we must consider that the islands are stressed due to a possible lattice mismatch between the island material and the substrate. The stressed islands relax at the boundary and exert a force on the substrate which is elastically distorted and mediates the interaction between islands during growth. Using elastic theory of continuous media, Marchenko [98] and others [99–101] could explain the spontaneous formation of periodic domain patterns in various systems. Mesoscopic domain ordering was observed on Si(100) [102], Au(111) [90], Cu(110) covered with oxygen [103] and Pd(110) covered with Cu [72]. The physical origin of these ordering phenomena are believed to be long range elastic interactions.

Let us consider the uniaxial problem. For a surface consisting of alternating stripes of two phases A and B of widths L_A and L_B respectively, the extra free energy per unit length ΔF due to the formation of domains is given by:

$$\Delta F = \frac{2F_s}{L_A + L_B} - \frac{2C_{el}}{L_A + L_B} \ln\left[\frac{L_A + L_B}{2\pi a}\sin(\pi\Theta)\right], \qquad (5.14)$$

$$\Theta = \frac{L_A}{L_A + L_B}; \quad 0 < \Theta < 1. \qquad (5.15)$$

ΔF is the sum of two terms: the first term is the free energy per unit length for the creation of a boundary, while the second term describes the elastic relaxation. The logarithm becomes infinity for the limits $\Theta = 0$ or 1 and it goes through a minimum for $\Theta = 0.5$. C_{el} depends on elastic parameters like the shear modulus and the Poisson ratio of the substrate as well as on the difference in the normal components of the surface stress between A and B domains, and "a" is the lattice constant. The equilibrium periodicity D is obtained by minimizing ΔF with respect to L_A, keeping Θ constant

$$D(\Theta) = \frac{\kappa}{\sin(\pi\Theta)} = L_A(\Theta) + L_B(\Theta), \qquad (5.16)$$

$$L_A(\Theta) = \frac{\kappa\Theta}{\sin(\pi\Theta)}, \qquad (5.17)$$

with $\kappa = 2\pi a \exp(1 + F_s/C_{el})$. L_A does not vary much as a function of Θ for $0 < \Theta < 0.7$, and the D curve as a function of Θ is also rather flat around a central value of about half coverage. It has been shown that there should exist a firm relation between the periodicity D at an intermediate coverage Θ and the width of a single island L_A at low coverage [101]. The ratio D/L_A is comprised between 0.25 and 0.33 for a coverage of about 0.5. This ratio is roughly independent of C_{el} and the free energy for the formation of one boundary.

Several results have been analyzed in terms of this theory. Island ordering has been observed on Cu/Pd(110) above a critical coverage, somewhere between 0.15 and 0.25 ML. At a coverage of 0.2 ML, the one-dimensional island-island correlation functions in the [001] direction shows a repeat distance of 40 Å. Since the average island width is 11 Å this leads to a ratio D/L_{Cu} of 0.28 which is within the predicted interval [72]. This interpretation of island ordering is strictly valid only for systems at equilibrium.

5.7.5 Organization on Vicinal Surfaces

When the mean free path Λ_a of adatoms on a surface (see Sect. 5.5.1) becomes larger than the typical width of a terrace Λ_t, the adatoms will reach the steps before they have the chance to meet another adatom on their path. As a result, in the so called "step flow" limit no islands form on the terraces anymore. In a favorable case, when the adatoms wet the steps, it leads to the formation of stripes of adatoms oriented parallel to the steps. A first trial of this approach has been attempted first on vicinal surfaces (surfaces with a slight miscut away from a dense crystallographic plane) that develop regularly spaced steps [104,105]. Results for Co/Cu(111) and Fe/Cu(111) are somewhat disappointing, since the stripes are irregular and sometimes segmented (see Fig. 5.22). As a result, structures are much less perfect than, for example, the Cu-stripes of Fig. 5.10b obtained by anisotropic diffusion. Linear arrays of Fe particles have been obtained by the shading technique [106] in which the metal vapor is deposited at a grazing incidence on a SiO coated NaCl grating.

5.7.6 Low Temperature Growth

As was mentioned in Sect. 5.5, low temperature deposition favors growth of small islands with a high surface density. This behavior manifests itself in a quite spectacular way in Co/Au(111) because of the large incidence on magnetism. As a matter of fact, Co/Au(111) is known to be a prototype of 3D growth since cobalt forms ordered bilayer clusters when grown at 300 K (Fig. 5.17a). However, as was shown by variable temperature STM [107],

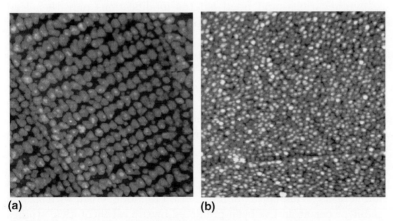

Fig. 5.17. Comparison between (**a**) 1.3 ML of Co grown on Au(111) at 300 K (Image 150 nm × 150 nm) and (**b**) 1.0 ML of Co grown on Au(111) at 30 K (Image 100 nm × 100 nm). In (**b**) compact stacking of monolayer clusters with a diameter of 0.5 nm occurs while the RT growth in (**a**) leads to bilayer clusters 7 nm in diameter

the nucleation is completely different when Co is grown on Au(111) at 30 K (Fig. 5.17b). Small monolayer thick islands with a high surface density form spontaneously [108], and a quasi layer-by-layer growth mode becomes apparent after completion of the first monolayer.

As was discussed previously, the growth of Co on Au(111) is normally governed by the nucleation on point dislocations of the zigzag reconstruction. This particular growth mode of bilayer-high islands is observed down to temperature as low as 150 K. At some point when lowering the substrate temperature, the mean free path of adatom diffusion becomes small compared to the separation between point dislocations of the zigzag reconstruction. The new islands density is then given by $N\ R^{1/3} \exp(-E_d/3kT)$, where R is the deposition rate and E_d the diffusion barrier for a cobalt adatom on the gold surface. A much higher density of islands than the one anticipated from self-assembly is therefore achieved in the early growth stage at low temperature (see Fig. 5.17). Growth, first proceeds by random nucleation of monolayer thick islands with lateral sizes of about 5 Å. The small monolayer thick islands then coalesce well before a critical size for bilayer islands is reached [107]. A quasi layer-by-layer, 2D growth, occurs contrary to the 3D growth observed at 300 K. As can be foreseen, the properties of films grown at low and high temperatures will be quite different as will be illustrated in the next section on the example of magnetism.

5.8 Magnetic Properties of Nanostructures

The magnetic properties of clusters deposited on a crystal surface or embedded in a matrix have become the subject of many interesting studies. The distortions, hybridizations, and bond lengths modifications may then lead to magnetic properties that are considerably different from those of the free clusters.

Since clusters (islands) on surfaces form the obliged pathway for the growth of thin films, in this part we will often refer to ultrathin magnetic films for which a large bulk of information exists. More details on surface magnetism, ultrathin magnetic structures and their magnetic properties can be found in [24,109]. While generally enhanced magnetic moments are expected at surfaces of "d" metals [110], or in thin films [111], they can be reduced in thin epitaxial films because of hybridization with the substrate, as for example in Fe/W [112]. Some metals that are non-ferromagnetic in the bulk, (e.g. V, Rh, Ru, Pd) are expected to become ferromagnetic in the form of free standing or ultrathin epitaxial films [113–115]. Magnetism with sizable magnetic moment per atom has been predicted by several theoretical studies for small clusters supported on silver [114,116] and for epitaxial ultrathin films deposited on noble metal [117]. However, at the present time, there is no clear experimental evidence of ferromagnetism in epitaxial layers or supported clusters of such metals.

For V, Rh or Ru clusters on Ag or Au, the situation remains controversial, but nearly all experiments give negative results [118–120,122,123]. Recent anomalous Hall effect and weak localization experiments might indicate that Ru atoms have a small magnetic moment when deposited on Pd at a very low coverage [123]. These results should however be confirmed by other magnetic characterization techniques. As demonstrated theoretically [124], the magnetism of these metals is highly sensitive to the local environment.

At surfaces or interfaces, the broken symmetry modifies the coupling and a specific surface anisotropy may appear, as suggested by Néel [125]. The surface anisotropy, which is actually also of magnetocrystalline origin, can either favor an easy magnetization direction parallel or perpendicular to the surface plane. It contains in a hidden way magneto-elastic effects occurring from the strain present at the surface or at interfaces [126]. Its energy contribution is: $E_s = SK_s$, where S is the surface area and K_s the surface anisotropy constant.

Another source of anisotropy is the shape anisotropy, arising from the long range dipolar interaction between the magnetic moments. It is strongly dependent on the object's shape (see e.g. [127]). The shape anisotropy may however become important for nanostructures with finite lateral sizes and high aspect ratia. The total anisotropy energy is given by the sum of all contributions. In a simple case it can be written in a form similar to equation (5.8). The equilibrium direction of the magnetization is obtained through minimizing the energy with respect to θ. A phase diagram providing the preferred directions as a function of K_2 and K_4 (see [128]) can be drawn.

In the following, we will analyze by means of several examples how ferromagnetism sets in as a function of adsorbate coverage. Here, we discuss the magnetic properties of nanostructures, from non-interacting well separated islands, to interconnected ones, forming stripes, to almost continuous films of two-dimensional character. Due to the important number of studies available, it is not possible to give a review of all systems studied. We limit ourselves to well characterized structures from both the magnetic and topographic point of views.

5.8.1 Isolated Clusters on Surfaces

From growth considerations (in Sect. 5.5), one expects the formation of separated and magnetically non-interacting clusters in the early stage of growth. Upon increasing the amount of deposited material, these clusters may coalesce to form bigger islands. At this stage the structures interact by exchange and/or dipolar coupling.

A typical illustration of this scenario is given by iron deposits on the W(110) surface. At 300 K, iron grows epitaxially in the form of monolayer thick islands on W(110) (Fig. 5.18). As long as the islands remain separated, up to an equivalent coverage of 0.58 ML of Fe, no ferromagnetic signal is

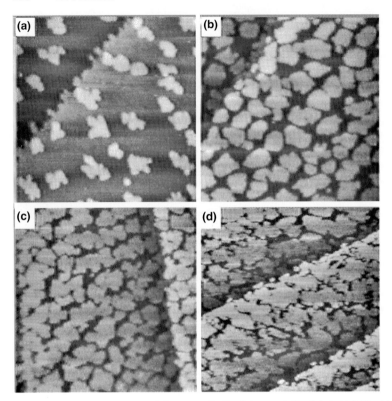

Fig. 5.18. STM image for (**a**) 0.23 ML Fe (**b**) 0.53 ML, (**c**) 0.66 ML, (**d**) 0.85 ML Fe grown on W(110) at room temperature. Images 70 nm × 70 nm [25]

observed as shown in Fig. 5.19 [25]. The sudden occurence of ferromagnetic order (Fig. 5.19), measured by Spin Polarized Low Energy Electron Diffraction (SPLEED), coincides with the topographic percolation observed for 0.6 ML.

Below 0.6 ML, the clusters are supposed to be in a superparamagnetic state; they are too small to possess a locked magnetic moment. This is a rather general behavior when a small amount of matter is deposited on the surface.

The same behavior can also be observed in a more controlled way when Co is deposited on the Au(111) surface. As mentioned earlier, Co grows in well ordered bilayer high clusters on the zigzag reconstruction of Au(111) (see e.g. Fig. 5.16). At room temperature, Co clusters nucleate at the elbows of the gold reconstruction and they expand laterally as a function of coverage, until they come close to contact at about 1.0 ML (since the clusters are bilayers, the gold surface is half-covered by Co at this stage). At coverages below 1.0 ML, where the assumption of non or weakly interacting clusters holds, the saturation fields needed to evidence the superparamagnetic state are so high, that *in situ* Kerr effect measurement on as-grown clusters are difficult. Therefore, the samples are measured *ex situ* in high fields by Kerr

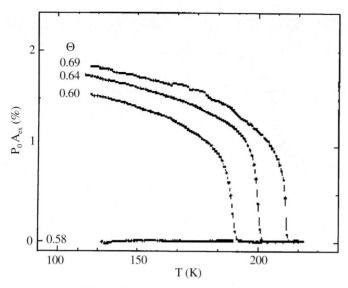

Fig. 5.19. SPLEED-Polarization as a function of measurement temperature for various Fe coverages grown at 300 K on W(110) [25]

effect or SQUID after having been covered by a protective layer. An anhysteretic magnetization curve, characteristic of the superparamagnetic state, is obtained (Fig. 5.20) [27,129,130]. The saturation fields are of the order of 1 Tesla.

Since the magnetization of superparamagnetic clusters is described by a Langevin function of argument $N\mu B/kBT$, the experimental curve allows, in principle, to determine the size of the clusters. One must assume that the clusters are all of the same size. This is a good approximation for self-organized Co clusters nucleated on the herringbone reconstruction of a single crystal. However, most magnetization curves [27,129] are measured on Co clusters deposited on thick gold films grown on mica substrates. The Au surface of such samples has more defects than a well prepared Au(111) single crystalline surface; therefore, the size distribution must be taken into account. Usually, a bimodal distribution allows to fit well the Langevin curve of an assembly of clusters [27].

Small and well calibrated clusters, like the Co clusters on Au(111), are suited to study electronic properties. As mentioned previously, modifications in the electronic structure can be expected in small clusters. Indeed, changes in the electronic structure are observed by Kerr spectroscopy below 2 ML Co coverage [131]. The consequence is an enhanced orbital magnetic moment, as evidenced by X-ray magnetic circular dichroïsm (Fig. 5.21).

The enhanced orbital contribution leads to a slightly larger anisotropy in the small clusters [132]. Similarly, a doubling of the orbital moment has been observed for small Fe clusters deposited on graphite [133]. A spectacular

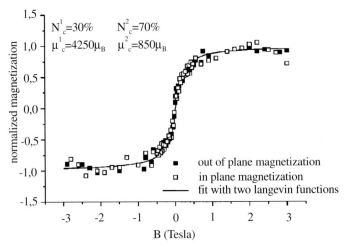

Fig. 5.20. SQUID magnetization curves recorded a 290 K for 0.4 ML Co on Au(111). The data points are fitted by a sum of two Langevin functions corresponding to two populations of clusters with different sizes (given are the values of the giant moments μ_c^i). The N_c^i are the percentages of either population [27]

increase of the magnetic moment has also been observed for Co and Fe atoms on Cs for extremely low amounts of deposited material [134].

Presently, such small organized dots are of great interest for the fundamental aspects of magnetism. However, because of their very low blocking temperature (below 30 K at about 0.5 ML) they are not yet suited for technological application in magnetic data storage. We show in the outlook of this review, how one might overcome this inconvenience.

5.8.2 Interacting Islands and Chains

Organized clusters, stripes, thin films allow to study how magnetic long range order appears and how the critical exponents evolve [109]. Monte Carlo methods allow one to simulate more realistic systems, by taking into account finite anisotropy terms [135,136] and the dipolar interaction [136,137] which both can stabilize long range order in 2D systems. These stabilizing effects are of great importance in low dimensional nanostructures. One must notice that the "magnetic" dimensionality is not always related to the real "topographic" dimension of the nanostructures.

Chains of Exchange Coupled Clusters. When the Co clusters on Au(111) (from Sect. 5.7.3) are allowed to grow laterally, they will come to close contact at an equivalent Co coverage of 1 ML. At this stage more or less continuous bilayer Co chains are formed. They are about 8 nm wide, and may be several hundred nanometers long. The chains, separated approximately by

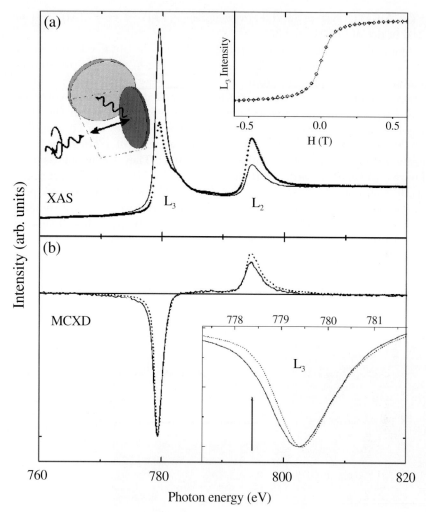

Fig. 5.21. (a) X-ray absorption spectra of the Co L-edges for 1.5 ML Co on Au(111) for the two photon spins in a 4 T field, at 20 K. (*Inset*: Langevin function fit of the magnetization curve). (b) XMCD difference spectra for 1.5 ML Co (*dashed line*) and 0.2 ML (*continuous line*). The differences are due to modifications in the orbital magnetic moment [132]

15 nm (see Fig. 5.26), remain superparamagnetic. They can be considered as unidimensional arrangements of fluctuating spin blocks with perpendicular anisotropy (1D Ising model) [27,129]. One does not expect any magnetic long range order since an Ising system looses its long range correlation at finite temperature. In this case, the magnetization curves are not supposed to fit a Langevin function [132], since the clusters are interacting with each other.

122   J.P. Bucher

The Co/Au(111) system in the coverage range $1\,\text{ML} < \Theta < 2\,\text{ML}$ will be treated in detail in Sect. 5.8.3.

As mentioned in Sect. 5.7.5, metallic stripes can be obtained upon deposition on vicinal surfaces. In the first growth stages, there is preferential nucleation of the deposited atoms at the step edges. The aggregation in the steps produces stripes whose width is controlled by the total coverage. The miscut angle of the crystal determines the spacing between the steps (hence the spacing between adjacent stripes). For example, the Fe stripes in Fig. 5.22 are obtained on a Cu(111) vicinal surface (miscut 1.2 deg) for 273 K deposition [105]. The stripes are aligned along [011]. They are monolayer-high and have a width of the order of 10 nm (coverage 0.3 ML). The chains are not perfectly continuous: it is seen that defects sometimes interrupt the stripes.

No magnetic longitudinal Kerr signal is obtained, irrespective of the field applied along or perpendicular to the stripes. Hysteretic polar Kerr loops are obtained at low temperatures, from 50 K for 0.3 ML to about 200 K for 0.8 ML (Fig. 5.23), suggesting a ferromagnetic phase for a quasi-1D system. However, the remanence of these Fe stripes is time-dependent. After applying a field pulse, the remanent magnetization decreases over a time of several seconds

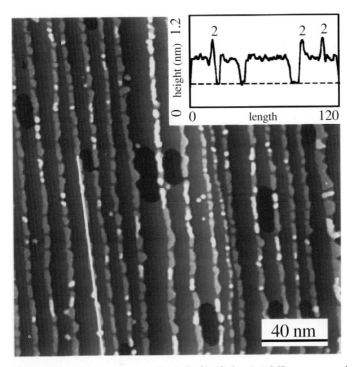

Fig. 5.22. Fe stripes on vicinal Cu(111) for 0.3 ML coverage A linescan is given along the white line [105]

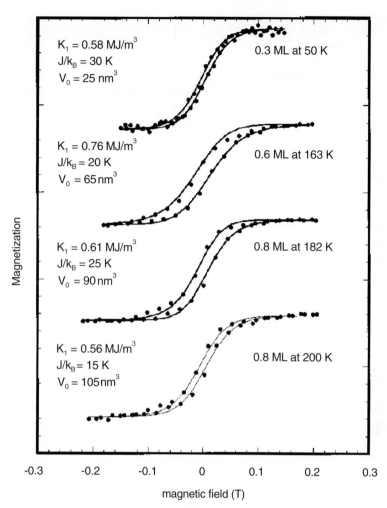

Fig. 5.23. Kerr loops of Fe on vicinal Cu(111) recorded for various temperatures and coverages. The solid lines are theoretical curves obtained in the Glauber-type model (see text) for the given anisotropies K_1, exchange constant J and volume V_0 [105]

(Fig. 5.24). This demonstrates the presence of fluctuating spin blocks, which can be partially frozen at low temperatures. Therefore, there is only magnetic long range order over a limited time scale. The hysteretic behavior comes from the fact that, at a given temperature, the sweeping rate of the field is much higher than the fluctuation time of the spin blocks. The magnetization curves can be fitted in a Ising model description, in which a Glauber-type dynamics has been introduced [105]. In this model, for a given field sweeping rate, the

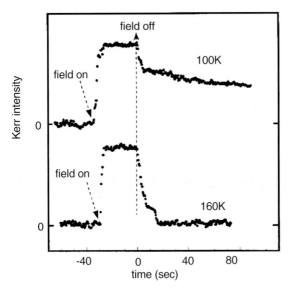

Fig. 5.24. Kerr signal as a function of time for an applied field pulse for a 0.8 ML Fe film on Cu(111) [105]

anisotropy, K_1, the volume V_0 of the spin blocks and the exchange coupling strength J between adjacent segments can be deduced from the fit.

Chains of Dipolar Coupled Clusters. Linear arrays of Fe particles were obtained by deposition of Fe vapor onto a SiO-coated NaCl(110) crystal (Fig. 5.25, right) [106]. Since Fe is deposited on a SiO buffer, the clusters are not expected to have a well defined cristallographic order. Therefore, and contrary to the Fe/W(110) case (below), the magnetic anisotropy should be weak. Once they have reached a critical size, the clusters within one array couple through the dipolar interaction and the magnetization aligns along the particle chains (Fig. 5.25 left and middle), for stray field minimization. The presence of long range magnetic order in this system could be reproduced by Monte Carlo simulations.

Dipolar Coupling Between Cluster Chains. Iron stripes were also obtained on a W(110) vicinal crystal (miscut 1.4 deg). In this case and contrary to the case of Fe on Cu(111), a persistent remanent magnetization is observed [138], although non-interacting single stripes should not present long range order. However, the Fe stripes are parallel to [001], whereas the easy magnetization axis of Fe/W(110) is along [1$\bar{1}$0], i.e. perpendicular to the Fe stripes because of a strong uniaxial in-plane surface anisotropy (despite the shape anisotropy). Since the spins are perpendicular to the stripes, the dipolar interaction across the stripes favors the alignment between spin blocks in two

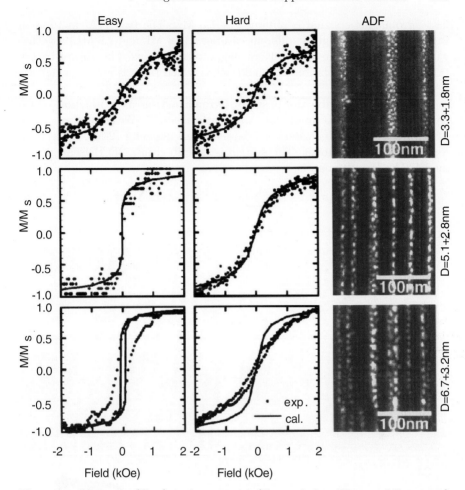

Fig. 5.25. Magnetic (Kerr) and structural (Transmission Electron Microscopy) data for Fe particles on SiO coated NaCl. The average particle size is given on the *right*. The easy magnetization axis is along the arrays. The continuous lines represent the Monte-Carlo simulated magnetization loops [106]

adjacent chains (to keep stray fields law). This additional dipolar interaction is expected to stabilize the long range ferromagnetic order.

5.8.3 The Two-Dimensional Limit

On some selected examples we illustrate now the transition from non, or weakly interacting clusters, towards correlated, two-dimensional structures.

Co Clusters on Au(111). We mentioned in Sect. 5.8.2 that at 1 ML coverage, the Co structures on Au(111) have no remanent magnetization. The first ferromagnetic signal is obtained at 1.6 ML. The two STM pictures in Fig. 5.26 show two slightly different coverages (1.4 ML and 1.6 ML) together with the corresponding in situ Kerr magnetization loops. The abrupt transition indicates a percolation threshold below which there is no ferromagnetic signal.

In order to describe the onset of ferromagnetism as a function of Co coverage, Monte Carlo simulations were performed [27]. We assume an $N \times N$ hexagonal array of Co clusters of 7.5 nm average diameter (2 ML thick) with a giant magnetic moment. The system is described in a 2D Heisenberg-like model. The Hamiltonian is:

$$H = \frac{\gamma}{4} \sum_{i,j} \delta_{ij} \sigma_i \varepsilon_i \sigma_j \varepsilon_j - KV \sum_i (\sigma_i^z \varepsilon_i)^2 - \mu_0 M H \sum_i \sigma_i^z \varepsilon_i , \qquad (5.18)$$

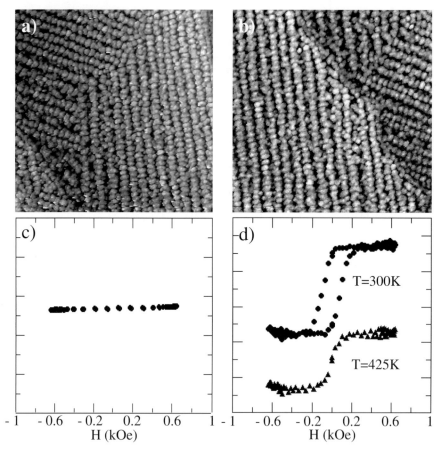

Fig. 5.26. STM images for **a)** 1.4 ML, **b)** 1.6 ML Co on Au(111) with the corresponding polar Kerr signals **c)** and **d)**. Images 200 nm × 200 nm [27]

where σ_i is the normalized magnetization of the cluster at a site i (M is the saturation magnetization), $\varepsilon_i = 1$ or 0 if the site is occupied or not by a cluster, $\delta_{ij} = 1$ for nearest neighbor clusters (0 otherwise), S is the contact surface between the clusters, V the individual cluster volume, K the perpendicular anisotropy constant, H the external field (applied in the z direction, perpendicular to the surface), γ is the wall energy, for two opposite magnetizations in two adjacent clusters. The wall energy can be estimated from a micromagnetic model ($\gamma = 2\sqrt{JK}$). The dipolar coupling between chains is neglected (this is legitimate at room temperature). The numerical values used for the simulation are reported in the caption of Fig. 5.27. The

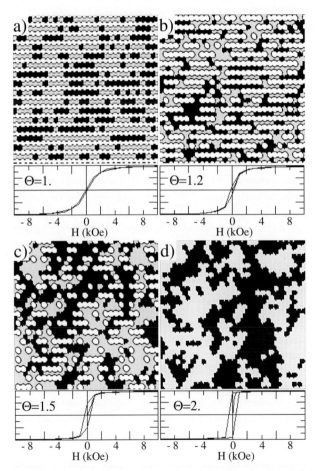

Fig. 5.27. Demagnetized σ_z maps of 30×30 sites for different coverages Θ. Unoccupied sites are in white, occupied sites with up (*down*) magnetization in grey (*black*). Hysteresis loops and first magnetization loops are shown [27]. $\gamma S/k_B = 125$ K, $KV/k_B = 1000$ K, $M/k_B = 1000$ K/Tesla

simulation starts at 1 ML coverage, i.e. from arrays of chains separated by lines of unoccupied sites (white in Fig. 5.27). Then the unoccupied sites are filled randomly with clusters. Figure 5.27 shows the σ_z-maps (projection of the magnetic moment on the z-axis) for several coverages in the demagnetized state at $T = 300$ K.

Below 1.5 ML (Fig. 5.27a and b) the up and down domains are small and confined within the chains. There are not enough connections to the adjacent row to allow a strong magnetic coupling. At about 1.5 ML, (Fig. 5.27c) the domain size increases and they start spreading over the whole surface. At 2 ML (Fig. 5.27d), the domain structure is very similar to the one observed experimentally [139]. The corresponding simulated magnetization curves are represented. One observes a drastic decrease of the saturation field from 1 to 1.5 ML. A significant remanent magnetization appears above 1.2 ML. These simulations are in quantitative agreement with the experimental results. Because of a perpendicular anisotropy, there is a transition from a 1D Ising-like system with no ferromagnetic long range order, towards a long-range ordered 2D Ising system.

Fe Clusters on W(110). As mentioned in Sect. 5.8.1, the abrupt onset of ferromagnetic order at 0.6 ML coverage corresponds to the percolation of the Fe islands as observed by STM. Since it is posible to obtain a thermodynamically stable and flat monolayer film, the Fe/W(110) system is an ideal system to study critical phenomena. The Curie temperature can be easily measured as well as critial exponents. It was shown that an Fe monolayer on W(110) can be described as a two-dimensional anisotropic Heisenberg system (actually the critical phenomena were analyzed on a Fe monolayer deposited at high temperature, which is thermodynamically stable) [140].

It is worth to mention that the room temperature grown films show a somewhat peculiar behavior in the 1.2 ML to 1.5 ML range. These Fe sesquilayers (i.e. one monolayer and a half) on W(110) loose their remanent magnetization [141] (Fig. 5.28). STM pictures show that in this coverage range double layer Fe islands are surrounded by an Fe monolayer sea [142].

Torsion Oscillatory Magnetometry coupled with Kerr effect measurements at different temperatures show that the Fe monolayer is ferromagnetic with an in-plane anisotropy, whereas the double layer islands are found to be superparamagnetic with a perpendicular anisotropy; some of them are blocked, while others not. The coupling between the double layer islands and the surrounding layer produces micromagnetic phenomena which is not very well understood as yet. The perpendicular anisotropy of the double layer islands is attributed to the high epitaxial strain in the Fe. Above 1.5 ML, misfit dislocations allow a strain release and the system returns to the normal situation of an in-plane magnetized film.

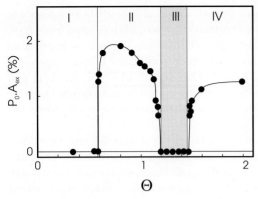

Fig. 5.28. SPLEED-Polarization as a function of Fe thickness. Note the absence of magnetic signal in region III [141]

Co Clusters on Cu(100). For low dimensional structures, interesting questions arise at the critical temperature above which long-range ferromagnetic order disappears. Is the critical temperature a true Curie temperature at which the spontaneous magnetization vanishes? Is it a superparamagnetic blocking temperature? Or, does the film just break into domains? These questions have been answered for the Co/Cu(100) system) [143]. A Co film, between 1 and 2 ML thick is ferromagnetic with a weak magnetocrystalline anisotropy; the magnetization is confined in-plane by the shape anisotropy. Kerr effect combined with SEMPA measurements showed that for such a thin Co film the spontaneous and remanent magnetization decrease in the same way with the temperature. This demonstrates that the critical temperature is a true Curie temperature. Above the Curie temperature, the system shows a strong susceptiblity in a weak external magnetic field, occurring from large fluctuating spin-blocks evidenced by SEMPA. The system behaves like a two-dimensional Heisenberg ferromagnet.

Influence of Growth Conditions. As mentioned in Sect. 5.5, when the density of island N_{sat}, (see (5.9)) is increased by reducing the growth temperature, the morphology of Co clusters grown on Au(111) is modified. When Co is deposited at 30 K on Au(111), small monolayer grains nucleate, coalesce and form rather compact Co layers (see Fig. 5.17b). These layers exhibit in plane ferromagnetism above 1.5 ML [107,108], contrary to room-temperature grown Co clusters which are magnetized perpendicularly between 1.5 and 5 ML. N_{sat} can also be enhanced by increasing the deposition rate R. However, since N_{sat} follows a power of R, whereas it is exponential with T, one has to increase R by several orders of magnitude to reach the same effect by cooling the substrate from 300 K to 30 K. Such elevated deposition rates are obtained by Pulsed Laser Deposition (PLD), for which the instantaneous flux is of the order of 10^3 to 10^4 ML/s. PLD-deposited Fe on Cu(100) has been shown to grow in a perfect layer-by-layer growth mode. These PLD-Fe layers are magnetized in-plane for coverages at which the thermal deposited films have out-of-plane anisotropy [144]. The exact origin of the in-plane anisotropy of Co and Fe when grown under these particular conditions is not completely clear yet. It seems that both low-temperature deposited Co on Au(111) and PLD-Fe on Cu(100) have a smaller magnetocrystalline anisotropy than for usual growth conditions and therefore the shape anisotropy can force the magnetization in-plane.

In the systems addressed above, the magnetic properties can be understood quite well in terms of percolating superparamagnetic islands. They can be described by a few classes of thermodynamic models, provided the growth and morphology are known accurately and provided one takes into account properly the anisotropy term and the dipolar coupling.

5.9 Conclusion and Outlook

A consistent although not yet complete picture has emerged from existing results. Experimentally, two extreme situations can be distinguished: (i) superparamagnetic relaxation which produces a narrow deflection profile due to statistical averaging, and (ii) locked moment behavior which produces a characteristic spreading of the deflection profile. The ferromagnetic transition metal clusters enter into the first category since their behaviors can be explained quite accurately by internal vibrational relaxation up to field of about 0.8 T. This effect only involves the clusters' vibrational temperature. In rare earth clusters, on the other hand, the dynamics of the clusters play an important role and both of the above mentioned situations are observed. We have shown that an adequate description of these clusters necessarily involves both, vibrational and rotational effects. Not only have both behaviors been identified, in agreement with theoretical predictions, but transitions from one behavior to the other under thermal activation have been observed in rare earth clusters.

The cluster study, by imposing different responses to the boundaries, reveals the completely different nature of ferromagnetism in 3d and 4f metals. The relatively simple situation of TM clusters, where the magnetic properties can be described by one single theory, contrasts with the complex situation of rare earth clusters where the appearance of magnetic magic numbers calls for an interpretation that includes the symmetry of the clusters. One of the major results in recent studies, is the discovery of magnetic relaxation effects for free clusters in the vacuum. It was also shown how fluctuations of the short range order alter the sharpness of the magnetic transition.

As far as supported clusters are concerned, the simultaneous analysis of STM topographs and magnetic properties of organized islands on surfaces has been conducted to improve an understanding of the interrelation between growth and magnetism. There is increasing evidence that particular properties of magnetic "films" appear as a result of the spontaneous nanostructuring during growth (interaction between spin block, pinning of magnetic walls on island boundaries, etc.) rather than due to intrinsic properties of monolithic, continuous films as sometimes assumed in the past. The recent development of variable temperature STM equipment allows the characterization of highly out of equilibrium materials with unexpected properties. This very fruitful approach also opens up new ways to exploit the atomic diffusion to tailor to a large extent the density and shapes of nanostructures. Although the feasibility of self-organization on dislocation networks has been demonstrated (Sect. 5.7), many problems remain to be addressed. A major challenge in future work will consist in tuning the periodicity of templates for self-organization purposes since presently only a few combinations of lattice mismatch of hetero-epitaxial systems (see Ag/Pt or Au(111)) are used. Ideally it should be possible to adjust continuously the periodicity from one value to another. In this respect, the possibility of using alloys of continuously varying concentration should certainly be considered. Another important possibility would be to use metal oxides as substrates as suggested by a recent work on MgO [145]. Finally, the possibility of growing the magnetic structures in the third dimension, in a fashion similar to what has been done on some semi-conducting systems (see for example [146]), should be investigated.

Up to now, ordered, well separated structures could be synthesized on a few systems. The magnetic blocking temperatures thus achieved for Co dots on gold only amount to a few tens of degrees Kelvin. Blocking temperatures in excess of 300 K are reached only when structures start to interact magnetically. In order to increase the blocking temperature of each dot separately, the volume of the dot must be increased; eventually its aspect ratio must be optimized. Along these lines, an attempt has been made to pile up self-organized cobalt dots on Au(111) [147].[1] Alternating gold and cobalt deposition under appropriate conditions, leads to arrays of 8 nm-high columns (4 nm in diameter). Although the self-organized columns obtained in this way have blocking temperatures close to 300 K, they nevertheless interact magnetically. Because of their strong magnetic anisotropies, small clusters of rare earth metals [9] or transition/rare earth metals [148] should also be considered for self-organization. It has been suggested that, owing to their high blocking temperature, mass selected Gd clusters from Gd_{20} to Gd_{100} could be used as storage media [149]. If the properties of these clusters remain unaltered by embedding them into an inert matrix, for instance SiO_2,

---

[1] By alternating gold and cobalt deposition under appropriate conditions, arrays of 8 nm-high columns (4 nm in diameter) are obtained. The Co bilayer dots, are covered with 4 ML Au at 425 K. Because of the different step heights of Co and Au, troughs are formed on top of the Co dots. Subsequent deposition of Co (at 425 K) leeds to preferential nucleation on top of the Co dots formed in the first place, an exchange between Co and Au is supposed to favor 3D growth of columns.

only a few tens of atoms would be required to store 1 bit of information as compared to 10^8 today.

Self-organized magnetic dots would provide a state of the art solution for large scale integration of very small particles for artificial atom devices, spin polarized transport structures, single electron Kondo resonance devices [2,150,151] which have been studied up to now only as single elemental devices. Finally self-organized magnetic dots embedded in a free-electron like matrix, are by construction, materials with very high density of interfaces. They are therefore particularly promising for giant magnetoresistance applications (GMR). Recent work on Co/Au/Co/Au(111) films [152] seems to confirm this prediction, although self-organization has not been verified in this work.

Acknowledgements. F. Scheurer and M. Romeo are greatly acknowledged for their help in preparing the figures.

References

1. P.S. Bechthold et al. Zeit. Phys. Chem. Neue Folge, **169**, 101 (1990).
2. R.C. Ashoori, H.L. Stormer, J.S. Weiner, L.N. Pfeiffer, K.W. Baldwin, K.W. West, Surf. Sci. **305**, 558 (1994).
3. L. Thomas et al. NATURE, **383**, 145 (1996).
4. P.J.M. van Bentum, R.T.M. Smokers and H. van Kempen, Phys. Rev. Lett. **60**, 2543 (1988).
5. Clusters of Atoms and Molecules I, Ed.: H. Haberland, Series in Chemical Physics, Vol. 52, Springer, 1994.
6. For an overview see the proceeding of the International Symposium on Cluster and Nanostructure Interfaces (ISCANI) held in Richmond 1999, Eds. R.S.P. Jena, S.N. Khanna and B.K. Rao (World Scientific, 2000).
7. W.A. de Heer, P. Milani and A. Chatelain, Phys. Rev. Lett. **65**, 488 (1990).
8. J.P. Bucher, D.C. Douglass and L.A. Bloomfield, Phys. Rev. Lett. **66**, 3052 (1991).
9. D.C. Douglass, J.P. Bucher and L.A. Bloomfield, Phys. Rev. Lett. **68**, 1774 (1992).
10. D.C. Douglass, A.J. Cox, J.P. Bucher and L.A. Bloomfield, Phys. Rev. B **47**, 12874 (1993).
11. I.M.L. Billas, J.A. Becker, A. Chatelain, W.A. de Heer, Phys. Rev. Lett. **71**, 4067 (1993).
12. A.J. Cox, J.G. Louderback and L.A. Bloomfield, Phys. Rev. Lett. 71, 923 (1993); A.J. Cox, J.G. Louderback, S.E. Apsel and L.A. Bloomfield, Phys. Rev. B **49**, 12295 (1994).
13. I.M.L. Billas, A. Chatelain, W.A. de Heer, Science **265**, 1682 (1994).
14. Physics Today, April 1995 special issue on "Magnetoelectronics".
15. B. Doudin and J.P. Ansermet, Europhys. News, **28**, 14 (1997).
16. D.D. Chambliss, R.J. Wilson and S. Chiang, J. Vac. Sci. Technol. B **9**, 993 (1991).
17. B. Voigtlander, G. Meyer, N.M. Amer, Phys. Rev. B **44**, 10354 (1991).

18. H. Brune, M. Giovannini, K. Bromann and K. Kern, NATURE, **394**, 451 (1998).
19. W. Gerlach and O. Stern, Z. Phys. 8, 110 (1992); ibid. **9**, 349 (1992).
20. F. Liu, M.R. Press, S.N. Khanna and Jena, Phys. Rev. B **42**, 976 (1990).
21. Phys. Rev. A **41**, 5691 (1991).
22. H. Roeder, E. Hahn, H. Brune, J.P. Bucher and K. Kern, NATURE, **366**, 141 (1993).
23. Z.Y. Zhang and M.G. Lagally, SCIENCE, **276**, 377 (1997).
24. Ultrathin Magnetic Structures I and II (Eds.: J.A.C. Bland, B. Heinrich), Springer, Berlin, Heidelberg, 1994.
25. H.J. Elmers, J. Hauschild, H. Höche, U. Gradmann, H. Bethge, D. Heuer, U. Köhler, Phys. Rev. Lett. **73**, 898 (1994).
26. S. Padovani, F. Scheurer and J.P. Bucher, Europhys. Lett. **45**, 327 (1999).
27. S. Padovani, I. Chado, F. Scheurer and J.P. Bucher, Phys. Rev. B **59**, 11887 (1999).
28. J.H. van Vleck, in *Theory of Electric and Magnetic Susceptibility* (Oxford Univ. Press, London, 1932).
29. D.C. Mattis, *The Theory of Magnetism* (Springer, Berlin, 1998).
30. V.L. Morruzi, J.F. Janak and A.R. Williams, *Calculated Electronic Properties of Metals* (Pergamon, New York, 1978).
31. J. Callaway and C.S. Wang, Phys. Rev. B **16**, 2095 (1977), ibid. **15**, 298 (1977).
32. Phys. Rev. B **19**, 2626 (1979).
33. C. Herring, in *Magnetism*, Vol. IV, Eds. G.T. Rado and H. Suhl (Academic Press, New York, 1966).
34. T. Kasuya, in *Magnetism*, Vol. IIB, Eds. G.T. Rado and H. Suhl (Academic Press, New York, 1966).
35. A.J. Freeman, in *Magnetic Properties of Rare Earth Metals*, Ed. R.J. Elliot (Plenum, London, 1972), pp. 245–333
36. A.R. Mackintosh, Physics Today, Europhys. News **19**, 41 (1988).
37. K.G. Wilson, Scientific American, August 1979.
38. L. Néel, Compt. Rend. Acad. Sci. **228**, 664 (1949); Ann. Geophys. **5**, 99 (1949).
39. I.S. Jacobs and C.P. Bean, in *Magnetism*, Vol. 3, Eds. G.T. Rado and H. Suhl (Academic Press, New York, 1963), pp. 271–350. see also E. Kneller, Handb. Phys. Bd. 18/2 (Springer, Berlin, 1966), p. 438.
40. S.E. Apsel, J.W. Emmert, J. Deng and L.A. Bloomfield, Phys. Rev. Lett. **76**, 1441 (1996).
41. F. Reuse and S.N. Khanna, Eur. J. Phys. D **6**, 77 (1999).
42. F. Reuse, S.N. Khanna and S. Bernel, Phys. Rev. B **52**, R11650 (1995).
43. S. Bouarab, A. Vega, M.J. Lopez, M.P. Iniguez, J.A. Alonso, Phys. Rev. B **55**, 13279 (1997).
44. F. Aguilera-Granja, S. Bouarab, M.J. Lopez, A. Vega, J.M. Montejano-Carrizales, M.P. Iniguez, J.A. Alonso, Phys. Rev. B **57**, 12469 (1998).
45. J.P. Bucher and L.A. Bloomfield, Phys. Rev. B **45**, 2537 (1992).
46. G.M. Pastor, J. Dorantes-Davila, Phys. Rev. B **52**, 13799 (1995).
47. D.M. Cox, et al. Phys. Rev. B **32**, 7290 (1985).
48. J.P. Bucher and L.A. Bloomfield, Int. J. Mod. Phys. B **7** 1079 (1993).
49. S.N. Khanna and S. Linderoth, Phys. Rev. Lett. **67**, 742 (1991).
50. B.V. Reddy, S.N. Khanna and B. Dunlap, Phys. Rev. Lett. **70**, 3323 (1993).

51. B.V. Reddy, S.K. Nayak, S.N. Khanna, B.K. Rao and P. Jena, Phys. Rev. B **59**, 5214 (1999); Y. Jinlong, F. Toigo and W. Kelin, Phys. Rev. B **50**, 7915 (1994); Z.Q. Li, J.Z. Yu, K. Ohno and Y. Kawazoe, J. Phys.: Cond. Matter **7**, 47 (1995); P. Villasenor-Gonzalez, J. Dorantes-Davila, H. Dreyssé, G.M. Pastor, Phys. Rev. B **55**, 15084 (1997); C. Barreteau, R. Guirado-Lopez, D. Spanjaard, M.C. Desjonqueres and A.M. Oles, Phys. Rev. B **61**, 7781 (2000).
52. D.C. Douglass, J.P. Bucher and L.A. Bloomfield, Phys. Rev. B **45**, 6341 (1992).
53. S.K. Sinha, Handbook of Physics and Chemistry of Rare Earths, Eds. K.A. Gschneidner and L.R. Eyring (North-Holland, Amsterdam, 1982), Vol. 1.
54. D.P. Pappas, A.P. Popov, A.N. Anisimov, B.V. Reddy, and S.N. Khanna, Phys. Rev. Lett. **76**, 4332 (1996).
55. D.C. Douglass, J.P. Bucher, D.B. Haynes and L.A. Bloomfield, in "From Clusters to Crystals", Eds. P. Jena, S.N. Khanna and B. Rao (Kluwer, Boston, 1992).
56. A. Einstein and W.J. de Haas, Verh. Dt. Phys. Ges. **17**, 152 (1915).
57. G.G. Scott, Rev. Mod. Phys. **34**, 102 (1962); see also: C. Kittel, Phys. Rev. **76**, 743 (1949).
58. G.H.O. Daalderop, P.J. Kelly and M.F.H. Schuurmans, in Ultrathin Magnetic Structures I (Eds.: J.A.C. Bland, B. Heinrich), Springer, Berlin, Heidelberg, 1994, p. 40.
59. G.M. Pastor, J. Dorantes-Davila, S. Pick and H. Dreyssé, Phys Rev. Lett. **75**, 326 (1995).
60. P. Jensen and K.H. Benneman, Z. Phys. D **26**, 246 (1993).
61. A. Maiti, L.M. Falicov, Phys. Rev. B **48**, 13596 (1993).
62. H. Roeder, H. Brune, J.P. Bucher and K. Kern, Surf. Sci. **298**, 121 (1993).
63. H. Brune et al. Surface Sci. **349**, L115–L122 (1996).
64. J. Venables, G.D.T. Spiller and M. Hanbücken, Rep. Prog. Phys. 47, 399 (1984).
65. H. Brune, Surface Sci. Rep. **31**, 121 (1998).
66. T.A. Witten and L.M. Sander, Phys. Rev. B **27**, 5686 (1983).
67. M.C. Bartlett and J.W. Evans, Surface Sci. **314**, L829 (1994).
68. T. Michely, M. Hohage, M. Bott and G. Comsa, Phys. Rev. Lett. **70**, 3943 (1993).
69. J. de la Figuera, J.E. Prieto, C. Ocal and R. Miranda, Surf. Sci. 307–309, 538–543 (1994) and Phys. Rev. B **47**, 13043 (1993).
70. Y.W. Mo, J. Kleiner, M.B. Webb, M. Lagally, Phys. Rev. Lett. **66**, 1998 (1991).
71. J.P. Bucher, E. Hahn, P. Fernandez, C. Massobrio and K. Kern, Europhys. Lett. **27**, 473 (1994).
72. E. Hahn, E. Kampshoff, A. Fricke, J.P. Bucher and K. Kern, Surf. Sci. **319**, 277 (1994).
73. A.A. Baski, J. Nogami and C.F. Quate, Phys. Rev. B **43**, 9316 (1991).
74. L.P. Nilsen et al. Phys. Rev. Lett. **71**, 754 (1993).
75. E. Bauer, Z. Krist. **110**, 372 (1958); E. Bauer and H. Poppa, Thin Solid Films, **12**, 167 (1972).
76. J.M. Blakely in *Introduction to the Properties of Crystal Surfaces*, Oxford, Pergamon (1973).

77. R. Kern, Bull. Miner. **101**, 202 (1978).
78. R. Kern, G. Lelay and J.J. Métois, Current Topics in Material Science Vol. 3, Eds. E. Kaldis, North Holland, Amsterdam (1979).
79. P. Müller and R. Kern, J. Cryst. Growth, **193**, 257 (1998).
80. L.Z. Mezay and J. Giber, Japn. J. Appl. Phys. **21**, 1569 (1982).
81. H.L. Skriver and N.M. Rosengaard, Phys. Rev. B **46**, 7157 (1992).
82. E. Bauer, Applications of Surface Science **11/12**, 479 (1982).
83. F. Gautier and D. Stoeffler, Surf. Sci. **249**, 265 (1991).
84. P. Müller and R. Kern, Microsc. Microanal. Microstruct. **8**, 229 (1997).
85. M.H. Grabow and G.H. Gilmer, Surf. Sci. **194**, 333 (1988).
86. R. Kern and P. Müller, Surf. Sci. **392**, 103 (1997); A. Bourret, Surf. Sci. **432**, 37 (1999).
87. R. Nötzel et al. Europhys. News, **27**, 148 (1996); see also J. Tersoff, C. Teichert and M.G. Lagally, Phys. Rev. Lett. **76**, 1675 (1996).
88. T. Kontorova and Y.I. Frenkel, Zh. Eksp. Teor. Fiz. **89**, 1340 (1938).
89. F.C. Frank and J.H. van der Merve, J. Proc. Roy. Soc. A **198**, 205 (1949).
90. J.V. Barth, H. Brune, G. Ertl and R.J. Behm, Phys. Rev. B **42**, 9307 (1990).
91. S. Narasimhan and D. Vanderbilt, Phys. Rev. Lett. **69**, 1564 (1992).
92. L.P. Nilsen et al. Phys. Rev. Lett. **71**, 754 (1993).
93. L. Zhang, J. van Elk and U. Diebold, Phys. Rev. B **57**, R4285 (1998).
94. I. Meunier, G. Tréglia, J.M. Gay, B. Aufray, B. Legrand, Phys. Rev. B **59**, 10910 (1999).
95. J.A. Strocio, D.T. Pierce, R.A. Dragosset, P.N. First, J. Vac. Sci. Technol. A **10**, 1981 (1992).
96. J.A. Meyer, I.D. Baike, E. Kopatzki, R.J. Behm, Surface Sci. **365**, L647 (1996).
97. J. Wollschläger and N.M. Amer, Surf. Sci. **277**, 1 (1992).
98. V.I. Marchenko, JETP Lett. **33**, 381 (1981).
99. O.L. Alerhand, D. Vanderbilt, R.D. Meade and J.D. Joannopoulos, Phys. Rev. Lett. **61**, 1973 (1988); D. Vanderbilt, in Computations for the Nano-scale, eds. P.E. Blöchl et al. (Kluwer, New York 1993) p. 1.
100. J. Tersoff and R.M. Tromp, Phys. Rev. Lett. **70**, 2782 (1993).
101. P. Zeppenfeld et al., Phys. Rev. Lett. **72**, 2737 (1994).
102. F.K. Men, W.E. Packard and M.B. Webb, Phys. Rev. Lett. **61**, 2469 (1988).
103. K. Kern et al. Phys. Rev. Lett. **67**, 855 (1991).
104. J. Camarero et al. Mat. Res. Symp. Proc. **384**, 49 (1995).
105. J. Shen, R. Skomski, M. Klaua, H. Jenniches, S. Sundar Manoharan, J. Kirschner, Phys. Rev. B **56**, 2340 (1997).
106. A. Sugawara and M.R. Scheinfein, Phys. Rev. B **56**, R8499 (1997).
107. I Chado, S. Padovani, F. Scheurer and J.P. Bucher, Appl. Surf. Sci. **164**, 42 (2000).
108. S. Padovani, F. Scheurer, I. Chado and J.P. Bucher, Phys. Rev. B **61**, 72 (2000).
109. U. Gradmann, in Handbook of Magnetic Materials Vol. 7 (Ed. K.H.J. Buschow), Elsevier (1993); H.C. Siegmann, J. Phys. Condens. Mat. **4**, 8385 (1992).
110. G. Allan, Surf. Sci. **74**, 79 (1978).
111. C.L. Fu, A.J. Freeman and T. Oguchi, Phys. Rev. Lett. **54**, 2700 (1985).
112. S.C. Hong, A.J. Freeman, C.L. Fu, Phys. Rev. B **38**, 12156 (1988).
113. S. Blügel, Phys. Rev. B **51**, 2025 (1995).
114. K. Wildberg, V.S. Stepanyuk, P. Lang, R. Zeller, P.H. Dederichs, Phys. Rev. Lett. **75**, 509 (1995).

115. H. Dreyssé, C. Demangeat, Surf. Sci. Rep. **28**, 65 (1997).
116. V.S. Stepanyuk, W. Hergert, P. Rennert, J. Izquierdo, A. Vega and L.C. Balbas, Phys. Rev. B **57**, R14020 (1998).
117. M.J. Zhu, D.M. Bylander and L. Kleinman, Phys. Rev. B **43**, 4007 (1991); O. Eriksson, R.C. Albers and A.M. Boring, Phys. Rev. Lett. **66**, 1350 (1991); S. Blügel, Phys. Rev. Lett. **68**, 851 (1992); Europhys. Lett. **18**, 257 (1992); R. Wu and A.J. Freeman, Phys. Rev. B **45**, 7222 (1992); S. Blügel, Solid State Comm. **84**, 621 (1992).
118. M. Stampanoni, A. Vaterlaus, D. Pescia, M. Aeschlimann, F. Meier, W. Dürr and S. Blügel, Phys. Rev. B **37**, 10380 (1988).
119. H. Li, S.C. Wu, D. Tian, Y.S. Li, J. Quinn and F. Jona, Phys. Rev. B **44**, 1438 (1991).
120. C. Liu and S.D. Bader, Phys. Rev. B **44**, 12062 (1991).
121. H. Beckmann, Fei Ye and G. Bergmann, Phys. Rev. Lett. **73**, 1715 (1994).
122. H. Beckmann and G. Bergmann, Phys. Rev. B **55**, 14350 (1997).
123. H. Beckmann and Bergmann, Eur. Phys. J. B **1**, 229 (1998).
124. I. Turek, J. Kudrnovsky, M. Sob, V. Drchal and P. Weinberger, Phys. Rev. Lett. **74**, 2551 (1995).
125. L. Néel, J. de Phys. Rad. **15**, 225 (1954).
126. D. Sander, Rep. Prog. Phys. **62**, 809 (1999).
127. D. Craik in *Magnetism*, Wiley, Chichester (1995).
128. Y. Millev and J. Kirschner, Phys. Rev. B **54**, 4137 (1996).
129. H. Takeshita, Y. Suzuki, H. Akinaga, W. Mitzutani, K. Ando, T. Katayama, A. Itoh, K. Tanaka, J. Magn. Magn. Mat. **165**, 38 (1997).
130. J. Xu, M.A. Howson, B.J. Hickey, D. Greig, E. Kolb, P. Veillet, N. Wiser, Phys. Rev. B **55**, 416 (1997).
131. H. Takeshita, Y. Suzuki, H. Akinaga, W. Mitzutani, K. Tanaka, T. Katayama, A. Itoh, Appl. Phys. Lett. **68**, 3040 (1996).
132. H.A. Dürr, S.S. Dhesi, E. Dudzik, D. Knabben, G. van der Laan, J.B. Goedkoop, F.U. Hillebrecht, Phys. Rev. B **59**, R701 (1999).
133. K.W. Edmonds, C. Binns, S.H. Baker, S.C. Thornton, C. Norris, J.B. Goedkoop, M. Finazzi and N. Brookes, Phys. Rev. B **60**, 472 (1999).
134. H. Beckmann and G. Bergmann, Phys. Rev. Lett. **83**, 2417 (1999).
135. R.P. Erickson and D.L. Mills, Phys. Rev. B **43**, 11527 (1991).
136. S.T. Chui, Phys. Rev. B **50**, 12559 (1994).
137. Y. Yafet J. Kwo and E.M. Gyorgy, Phys Rev. B **33**, 6519 (1986).
138. J. Hauschild, H.J. Elmers, U. Gradmann, Phys. Rev. B **57**, R677 (1998).
139. R. Allenspach, M. Stampanoni, A. Bischof, Phys. Rev. Lett. **65**, 3344 (1990).
140. H.J. Elmers, J. Hauschild and U. Gradmann, Phys. Rev. B. **54**, 15224 (1996).
141. H.J. Elmers, J. Hauschild, H. Fritzsche, G. Liu and U. Gradmann, Phys. Rev. Lett. **75**, 2031 (1995).
142. N. Weber, K. Wagner, H.J. Elmers, J. Hauschild, U. Gradmann, Phys. Rev. B **55**, 14121 (1997).
143. D. Kerkmann, D. Pescia, R. Allenspach, Phys. Rev. Lett. **68**, 686 (1992).
144. H. Jenniches, J. Shen, Ch.V. Mohan, S. Sundar Manoharan, J. Barthel, P. Ohresser, M. Klaua and J. Kirschner, Phys. Rev. B **59**, 1196 (1999).
145. Noguera et al. in Chemisorption and reactivity on Supported Clusters an Thin Films. (Kluver Ed. 1997).
146. Q. Xie, A. Madhukar, P. Chen and N. Kobayashi, Phys. Rev. Lett. **75**, 2542 (1995).

147. O. Fruchart, M. Klaua, J. Barthel and J. Kirschner, Phys. Rev. Lett. **83**, 2769 (1999).
148. A. Mougin, C. Dufour, K. Dumesnil, N. Maloufi and Ph. Mangin, Phys. Rev. B **59**, 5950 (1999).
149. United States Patent on "High density magnetic recording medium" Patent Number: 5,830,588. Nov. 3, 1998. Inventors: D.C. Douglass, J.P. Bucher, L.A. Bloomfield.
150. B. Doudin, G. Redmond, S.E. Gilbert, J.-Ph. Ansermet, Phys. Rev. Lett. **79**, 933 (1997).
151. V. Madhavan, W. Chen, T. Jamneala, M.F. Crommie, N.S. Wingreen, Science **280**, 567 (1998).
152. J. Corno, M. Galtier, D. Renard, J.P. Renard and F. Trigui, Eur. Phys. J. B **10**, 223 (1999).

6 Magnetism in Free Clusters and in $Mn_{12}O_{12}$-Acetate Nanomagnets

S.N. Khanna, C. Ashman, M.R. Pederson and J. Kortus

6.1 Introduction

Novel properties and behaviors have been found to emerge as the size of the material is reduced from the bulk to the nanometer or sub-nanometer regime [1]. These include atomic clusters containing ten to several thousand atoms, nanoscale materials, mono- and multi-layers, clusters deposited on surfaces, and the nanocrystal superlattices. The atomic arrangements and the physical, chemical, electronic, and the magnetic properties of these new materials can be vastly different from those of the individual atoms/molecules or the solids. This novelty arises not by a shear reduction in size, but by the occurrence of an entirely new class of phenomenon inherent to or becoming dominant at the reduced size. Further, the properties evolve with size, dimension and the composition. These developments have provided hope that novel materials with entirely new properties could be synthesized. This new era of nano-technology is expected to revolutionize the science and technology in the 21st century. It is, however, clear that the renewed promise rests on our ability to fundamentally understand the behavior of matter at small sizes or dimensions and to produce these materials in large quantities. It is important to note that for practical applications, the clusters or nano-particles have to be assembled to form nano-composites, deposited on substrates, or made to assemble in superlattices. A fundamental understanding of the role of interface is therefore critical to any applications. An area that has potential for tremendous applications and is likely to be most affected by the new technology is the area of magnetism. It is one of the ancient fields of science. However, despite many years of study to understand the origin of magnetic order, the interest in the field has not diminished. This is due to challenge of including relativistic and many body effects in the theory as well as due to practical technological applications ranging from magnetic recording media to motors, transformers, credit cards, engines, radio and television, microwave devices, xerographic copiers, magnetic refrigerators, and permanent magnets. There is thus a constant search for new magnetic materials with novel properties. As the pace of technology carries us to smaller and smaller devices, attention is being focussed on new effects such as magnetic order in clusters and nano-particles, oscillatory exchange coupling in multi layers, giant magneto-resistance (GMR) in magnetic multi layers, GMR in granular alloys composed of magnetic grains or clusters, and recent interest in resonant quantum tunneling in nano-magnets. These developments will have a strong impact on the future magnetic devices. An understanding of the factors controlling the magnetic order at the nano- and sub-nano scale is essential to develop these devices. The present review focusses on the space quantization and other quantum effects in free and embedded clusters. We first review free clusters and why the space quantization has not been observed in the Stern-Gerlach experiments on these systems. We then discuss the effect of adsorbates on the magnetic properties and briefly discuss an intriguing electronic quantum effect where the adsorption of successive H atoms leads to an oscillatory change in the magnetic moment of Ni_n clusters. Finally, we come back to space quantization and discuss its observation in $Mn_{12}O_{12}$-acetate and octonuclear iron oxo-hydroxo Fe_8 nanomagnets. The main focus will be on free and passivated $Mn_{12}O_{12}$ clusters where we will present results of electronic structure studies within the density functional theory. It is shown that the ground state of a free $Mn_{12}O_{12}$ cluster is an antiferromagnetic tower-like structure with no net moment. Electronic structure studies on a $(Mn_{12}O_{12}(RCOO)_{16}(H_2O)_4)$ passivated cluster as a model of the bulk nanomagnet show that the non-magnetic host plays a critical role in stabilizing the D_2 geometric structure which is responsible for a ferrimagnetic $Mn_{12}O_{12}$ core. The studies enable us to estimate the local magnetic moments at the various sites and provide insight into the nature of the spin tunneling. The highlight of the work is a new approach to numerically calculate the second-order magnetic anisotropy energy. It is based on a simplified but exact method for the calculation of spin-orbit coupling in multicenter systems, is free of shape approximations, and is independent of the choice of basis sets. Using this approach, we provide an analysis of various contributions to this quantity.

6.2 Magnetic Moment of Free Clusters in Beams

Extensive theoretical and experimental work on clusters and nano-magnets over the past decade has shown that the magnetic moment as well as the magnetic ordering can be substantially different at small sizes [2]. The earlier work on the magnetic properties of clusters was motivated by the expectation that the reduction in size would lead to enhanced magnetic moments. Using an electronic structure scheme based upon a combination of the molecular clusters and tight binding formulation, Liu et al. [3] examined the variation of the magnetic moment of Fe, Co, and Ni as one goes from solid to atoms by examining magnetic moment on atoms corresponding to crystallographic planes of various orientations, atoms in isolated planes, and chains. The local coordination in these wide ranges of environments varies from 2 to as much as 14. Results indicate that the magnetic moment is primarily sensitive to the local nearest neighbor coordination and as the coordination number is reduced, the magnetic moment is found to increase from the bulk value to that in an isolated atom. The increase in moment is primarily due to the narrowing of the energy bands with coordination, suggesting that the clusters of these solids would be more magnetic than their bulk counterparts.

De Heer et al. [4,5] were the first to carry out a systematic study of the magnetic moment of Fe_n clusters. They used the Stern-Gerlach experiment that is generally used to observe space quantization in atoms. It was expected that the beam containing clusters would split in to $2J+1$ components where J is the total angular momentum quantum number as in case of atoms. The results were quite surprising. All the clusters deviated along the same direction. Further, the measured magnetic moments per atom calculated using the deflections were far below the corresponding bulk values and increased with size. Following their work, Bloomfield et al. [6] carried out experiments on Co_n clusters. They found similar results, but also observed that the moment per atom increased linearly with size. These findings were quite startling as all available theories predicted an enhanced magnetic moment at the small sizes.

The apparent disagreement between the experiment and the existing theory was almost immediately resolved by Khanna and Linderoth [7]. They suggested that since the cluster sizes are much smaller than the typical size of magnetic domains, the atomic spins are exchange coupled and the cluster behaves like a giant magnet with a combined moment from individual atoms. They argued, however, that the reduction in size leads to a reduction in anisotropy energy and for Fe or Co clusters containing 10 to few hundred atoms, the anisotropy energy is much smaller than the thermal energy at the cluster temperatures. While, the local moments in these small clusters were ferromagnetically coupled, they pointed out that the overall cluster moment changes direction under thermal fluctuations and that the clusters undergo super-paramagnetic (SP) relaxation. In a magnetic field B, the total magnetic moment of the monodomain cluster, $N\mu$, where μ is the moment per atom, tends to align with the magnetic field. Thermal fluctuations, however, tend to counter this alignment. The key thing to note is that the typical superparamagnetic relaxation time is 10^{-10}–10^{-12} s whereas the passage time through the magnet are around 1–1000 μs. Hence, in actual experiments, one measures the average magnetization, M, that is related to the intrinsic moment via the relation,

$$M_{\mathrm{avg}} = N\mu \left[\coth\left(\frac{N\mu B}{k_\mathrm{B} T}\right) - \frac{k_\mathrm{B} T}{N\mu B} \right], \tag{6.1}$$

where T is the temperature of the cluster. Note that since each cluster spans the phase space during its passage, they all exhibit the same average magnetization and hence all the clusters are deviated in the same direction. The uniform deflection does, however, raise an interesting dilemma. It is well known that for single atoms, the Stern Gerlach profile corresponds to $2J+1$ multiplets where J is the total angular momentum quantum number. Why is it then that the cluster beam does not exhibit the multiplet structure?

The reason lies in the conservation of angular momentum. In both cases, the presence of the external field causes $2J+1$ fold equally spaced energy values. However, in case of atoms, the conservation of the angular momentum prevents any relaxation of the states and the beam splits into $2J+1$ components. In case of clusters, in addition to the electronic angular momentum, the cluster has an angular momentum associated with its own rotation. It can be shown that a coupling between the electronic and the massive angular momentum is present due to an effective spin orbit like interaction. This can be seen by transforming to a reference frame attached to an electron where the ions rotate around the axis. It is this coupling that allows the relaxation in clusters provided the thermal energy is larger than the anisotropy energy. For cluster sizes of interest, $N\mu_B$ is less than $k_B T$ and the 6.1 reduces to

$$M_{\text{avg}} = N^2 \frac{\mu^2 B}{3k_B T}, \tag{6.2}$$

that corresponds to an observed effective moment, μ_{eff}, of

$$\mu_{\text{eff}} = N \frac{\mu^2 B}{3k_B T} \tag{6.3}$$

in agreement with the experiments that show that the observed moment increased linearly with the size of the cluster.

The above model allows one to extract the true cluster moment per atom, μ, from the measured μ_{eff} knowing the cluster temperature. As discussed in the chapter by J.P. Bucher, the model is now routinely used by experimentalists to analyze the experimental data and it is found that the bulk magnetic moment is enhanced by almost 35% in small clusters.

Although the SP model has provided an excellent framework to interpret the experimental data, there are limitations one must be aware of. The model assumes that the anisotropy energy is small compared to thermal energy. While this is a good first approximation, the shape, surface, and volume contributions to anisotropy can become significant at small sizes. For example, in the case of granular alloys of Fe in matrices, Xiao et al. [8] have measured a value of anisotropy energy per unit volume to be 2×10^7 ergs/cm^3 for 20–40 Å particles. Using the same value for clusters, one finds that Fe$_n$ clusters having 120–140 atoms will have an anisotropy energy equivalent to a thermal energy at around 90 K. While there is no such measured data on nickel particles, one can expect a similar trend for anisotropy energy. As shown by Jensen and Bennemann [9], the inclusion of anisotropy contributions can have a quantitative effect on the moments derived using the super-paramagnetic model. In addition, one needs to know the precise temperature of the cluster to derive its intrinsic magnetic moment from the super-paramagnetic model. Determination of the cluster temperature has remained a controversial issue amongst experimentalists. Since there is no unambiguous method to derive the cluster temperature, the experimental magnetic moments obtained using the super-paramagnetic model have to be regarded as tentative.

6.3 Oscillatory Change in the Magnetic Moment of Ni_n Clusters upon H Adsorption

While free clusters are interesting, for applications the clusters have to be deposited on surfaces or coated with ligands. Consequently, one of the important problems is the effect of adsorbates on the properties of clusters. Here we discuss a provocative new effect recently reported by the current authors concerning the chemisorption of hydrogen around Ni clusters. It is shown that unlike bulk Ni, the hydrogen chemisorption, in most clusters, leads to an oscillatory change of the magnetic moment when successive H atoms are added. A detailed analysis of the electronic structure reveals that the changes in the magnetic moment can be related to the relative position of the lowest unoccupied molecular orbital (LUMO) of the majority and minority spin electrons.

The investigations of Ashman et al. focussed on Ni_nH_m clusters containing up to 4 Ni and 2 H atoms. Their studies were carried out using the NRLMOL set of codes developed by Pederson and co-workers over a number of years [10,11]. These codes are based on density functional approach and use the generalized gradient approximation for the exchange and correlation. Since the details of the method are published in a separate publication, here we briefly review the results. Figure 6.1 shows the ground state geometry of the Ni_mH_m clusters. For NiH_2, the ground state corresponds to a H–H distance of 1.74 Å indicating that the H is absorbed dissociatively as is the case on bulk surfaces. For Ni_2, the first H can be absorbed in an on top or a bridge location. One finds that the on-top site is unstable and the atom moves to the bridge site. For Ni_2H_2, the two configurations corresponding to spin singlet and triplet states are almost degenerate (only triplet configuration is shown in Fig.6.1). In both situations the H-H distance is more than 3 Å again indicating that the H is absorbed dissociatively. Ni_3 is the smallest cluster where H could occupy an on-top, bridge, or a hollow site. The studies indicated that the bridge site was more stable than the on-top and hollow site by 0.13 and 0.16 eV respectively. The next H is also absorbed at a bridge site and decorates the other Ni–Ni bond. This behavior continues to Ni_4 that has a D_{2d} structure. The first and the second H are absorbed at the adjacent bridge sites. It is interesting to note that in all cases, the absorbed species does not lead to a rearrangement of the basic cluster. The Ni–Ni bond lengths are almost the same as in pure clusters. The Ni–H bond lengths do increase with cluster size and coverage from 1.47 Å in NiH to 1.64 Å in Ni_4H_2.

In addition to the geometries, we give in Table 6.1 the binding energy of the various pure clusters and the gain in energy ΔE_H in adding the H atom defined by

$$\Delta E_H = -[E(Ni_nH_x) - E(Ni_nH_{x-1}) - E(H)] \ . \tag{6.4}$$

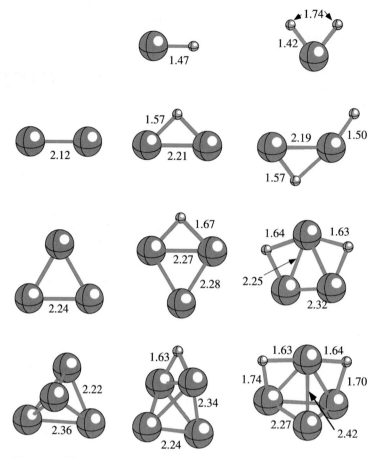

Fig. 6.1. The ground state geometries of the Ni_mH_n clusters ($m = 1-4$, $n = 0, 2$)

For pure clusters the binding energy, ΔE, is defined as

$$\Delta E = -[E(\mathrm{Ni}_n) - nE(\mathrm{Ni})] \,. \tag{6.5}$$

One of the most unusual effects of the coverage is seen in the magnetic moment. Table 6.1 lists the spin magnetic moment of pure and hydrogenated clusters. Except for a single Ni atom, addition of successive H atoms leads to an oscillatory change in the spin magnetic moment of clusters. In Ni_2, and Ni_4, the first H decreases the spin magnetic moment while the second H enhances the magnetic moment. In Ni_3, on the other hand, the first H increases the moment while the additional H decreases the moment. All this is particularly interesting since H absorption on Ni surfaces is known to always quench the magnetic moment. An analysis of the electronic structure of the resulting clusters shows that the H induced states are almost 4.8 eV below

Table 6.1. Magnetic moment (μ_B) and binding energy (eV) of Ni_n and Ni_nH_m Clusters. For pure clusters, the binding energy ΔE was calculated using (6.5). For Ni_nH_m clusters, the binding energy ΔE_H was calculated using (6.4)

Cluster	Mag. Mom. μ_B	ΔEnergy (eV)
NiH	1	2.92
NiH_2	0	2.59
Ni_2	2	2.76
Ni_2H	1	2.71
Ni_2H_2	2	2.58
Ni_3	2	5.40
Ni_3H	3	2.70
Ni_3H_2	2	2.92
Ni_4	4	8.39
Ni_4H	3	2.93
Ni_4H_2	4	2.82

the Fermi energy. How does then the H lead to an oscillatory progression in the magnetic moment?

The changes in magnetic moment of a Ni_nH_m can be understood within a simple model discussed by Fournier and Salahub [12] for magnetization of transition metal surfaces. Consider a filled molecular orbital of Ni_nH_{m-1} cluster with a pair of electrons interacting with the H 1s state. The interaction leads to the formation of a low lying bonding and a high-energy anti-bonding molecular orbital. Of the three electrons involved in the process, the two occupy the bonding orbital. The third electron goes to the LUMO of Ni_nH_{m-1} orbital. The change in moment is thus related to the location of the lowest unoccupied orbital of the preceding cluster. A similar criterion can also be obtained using another point [12] of view. One can consider the H atom as a proton and an electron. The additional electron occupies the spin state with lowest LUMO while the proton is screened by the d-states of the neighboring Ni sites. Note that this is not the only consideration. If the LUMO of the preceding cluster belongs to the minority manifold and the LUMO of majority is only slightly higher, the additional electron may still go to majority manifold as the exchange coupling could lead to a rearrangement of the manifolds. In brief, it is the difference, δE, between the LUMO of the majority and the minority spin manifolds that controls the change in moment. For the cases where this quantity is positive, the moment is expected to increase. When this quantity is highly negative, the magnetic moment will decrease. Table 6.2 shows the HOMO and LUMO of all the clusters. It is seen that when δE is less than -0.40 eV, the spin magnetic moment does decrease upon addition of H! For Ni_2H_2, their studies show the existence of a singlet isomer and if one includes it, the trend is observed here too.

Table 6.2. The HOMO and LUMO levels (eV) of the majority and minority spin states and δE in Ni_n and $Ni_n H_m$ Clusters

Cluster	Majority		Minority		δE
	HOMO	LUMO	HOMO	LUMO	eV
Ni	−0.17448	−0.03598	−0.14358	−0.14327	−2.92
NiH	−0.19549	−0.10286	−0.17134	−0.17089	−1.85
NiH_2	−0.15474	−0.10239	−0.15474	−0.10239	0.00
Ni_2	−0.16665	−0.11089	−0.14269	−0.13164	−0.56
Ni_2H	−0.14291	−0.10442	−0.13003	−0.11772	−0.36
Ni_2H_2	−0.18847	−0.12347	−0.17514	−0.15385	−0.84
Ni_3	−0.14337	−0.107217	−0.12351	−0.111745	−0.28
Ni_3H	−0.15314	−0.12253	−0.14414	−0.13944	−0.46
Ni_3H_2	−0.16005	−0.12783	−0.14273	−0.13234	0.12
Ni_4	−0.13081	−0.12112	−0.14812	−0.12203	−0.02
Ni_4H	−0.14602	−0.11872	−0.13013	−0.12218	−0.09
Ni_4H_2	−0.15828	−0.12725	−0.15277	−0.14245	−0.41

To summarize this section, the absorbates can produce interesting magnetic effects in clusters. For the H adsorption, the behavior is rooted in the electronic structure of the preceding cluster and the changes in the magnetic moment are indicative of the relative ordering of the majority and minority LUMO's. In the above case, the absorbate does not lead to a drastic change in the geometry of cluster. We now discuss a case where the ligands can dramatically affect the shape of the cluster and lead to new magnetic effects.

6.4 Quantum Tunneling and Atomic, Electronic and Magnetic Structure of $Mn_{12}O_{12}$-Acetate

As mentioned above, one of the most notable quantum phenomena associated with spin or magnetic quantum number is the space quantization generally observed in atoms via Stern-Gerlach experiments. As mentioned above, while the Stern-Gerlach setup [4] exhibit space quantization in atoms, in the case of clusters, the reduction in anisotropy leads to superparamagnetic relaxation. Can one observe space quantization in clusters?

In principle, the space quantization could be observed via macroscopic magnetization experiments provided: (1) One can generate size selected magnetic clusters in macroscopic quantities. (2) Such clusters could be assembled in an ordered array in such a way that the interactions between individual clusters are small enough not to alter the quantized levels. (3) Thermal and other fluctuations could be reduced. (4) There is a finite anisotropy which separates the various multiplets. (5) The angular momentum from individual clusters could be absorbed to permit transitions between various quantized directions.

Interestingly, it has been possible to make crystals containing magnetic clusters that satisfy all the above requirements. It has been shown that the space quantization can be observed in macroscopic magnetic hysterisis loops of cluster solids like $Mn_{12}O_{12}$-Acetate [13] and octonuclear iron oxo-hydroxo Fe_8 nanomagnets [14]. These solids contains $Mn_{12}O_{12}$ and octanuclear iron (III) oxo-hydroxo Fe_8 clusters which are magnetic, but different clusters interact only weakly via the dipolar fields. In this section we mainly focus on $Mn_{12}O_{12}$-acetate. At ordinary temperatures, the clusters undergo superparamagnetic relaxation. At low temperatures, however, experiments indicate that as one reduces the magnetic field from its saturation value, at fields below zero, the magnetization exhibits steps with decreasing field. Steps are also observed when one starts with a negative saturation field and goes to positive values. The magnetic relaxation rates are enhanced at the fields corresponding to the steps. These effects are proposed to be associated with resonant quantum tunneling of magnetization (QTM) [15]. This observation of quantum effects at a macroscopic scale has generated considerable excitement and the recent experiments indicate that the system exhibits different relaxation regimes [16].

The Mn-Acetate, discovered by Lis [17], is composed of acetate radicals, water and acetic acid molecules with a formula of
$[Mn_{12}(CH_3COO)_{16}(H_2O)_4O_{12}]:2(CH_3COOH):4(H_2O)$. The acetate contains $Mn_{12}O_{12}$ clusters, each carrying a net moment of $20.0\,\mu_B$. The individual clusters are separated by 15 Å of space filled with the non-magnetic host. The magnetic interactions between the $Mn_{12}O_{12}$ clusters are therefore weak and dipolar in nature. The molecule has an overall tetragonal symmetry and each $Mn_{12}O_{12}$ cluster is oriented along an S_4 easy axis. Further, the nanomagnet has an anisotropy energy of uniaxial symmetry. The individual $Mn_{12}O_{12}$ clusters with a total spin of 10 can then assume 21 $(2S+1)$ orientations along the Z-axis and the energies depend on the azimuthal quantum number m. In the absence of the external field, the states with $+m$ or $-m$ quantum numbers are degenerate and the spectrum consists of 11 discrete energies. Application of an external magnetic field along the z-direction, increases the binding energy of the parallel azimuthal magnetic $(+m)$ states while it reduces those of the $(-m)$ states. It has been proposed that the steps correspond to the field values such that the energy of an m state coincides with that of a $-m'$ state. At these values, the relaxation rate increases and a resonant quantum tunneling [18] of magnetization (QTM) takes place. Further evidence for the QTM has come from a variety of other experiments including the electron paramagnetic resonance (EPR) [19]. There is considerable evidence that the tunneling is assisted by the surrounding host indicating that the nonmagnetic passivators play a critical role in the quantum phenomenon.

The $Mn_{12}O_{12}$ clusters in the acetate host have a D_2 structure shown in Fig. 6.2a. The central core of the cluster consists of a tetrahedron of four Mn and four O atoms bonded to acetate radicals. It is surrounded by an outer

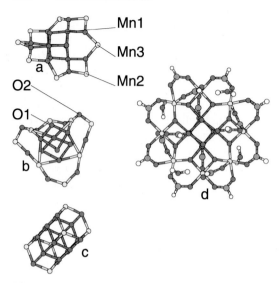

Fig. 6.2. Pictures of molecules studied in this work. In the left panel are three different $Mn_{12}O_{12}$ geometries. The righthand panel (**d**) shows the geometry of the relaxed $Mn_{12}O_{12}(RCOO)_{16}(H_2O)_4$ cluster and the bond lengths and angles are essentially identical to the experimental geometry determined from the X-ray data. (**a**) shows the $Mn_{12}O_{12}$ core of this molecule. With the absence of the RCOO and H_2O ligands, the D_2 $Mn_{12}O_{12}$ core relaxes to the D_{2d} structure shown in (**b**). We find that the Castleman tower (**c**) is locally stable and more stable than the D_{2d} core (**b**) by 5.14 eV. Structures a–c are nominally composed of Mn^{+2} and O^{-2} ions with local moments of $5\,\mu_B$ on the Mn atoms. However structures a–b exhibit ferrimagnetic ordering with total moments of $20\,\mu_B$ while the Castleman tower exhibits antiferromagnetic ordering with no net moment. The ligated structure (**d**) also exhibits ferrimagnetic ordering but the charge states of the inner and outer are nominally +4 and +3 respectively leading to moments of $-3\,\mu_B$ on the inner Mn atoms and $+4\,\mu_B$ on the outer Mn atoms. The Mn of different spins are shaded dark and light, respectively

shell of eight Mn and eight O atoms. The magnetic $Mn_{12}O_{12}$ cluster is bonded to CH_3COO acetate radicals and water molecules. The oxygen and acetate radicals are proposed to be in a charge states of -2 and -1 respectively. Further, the inner core four Mn (marked as Mn_1 in Fig. 6.2a) are in a Mn^{4+} state and carry a localized moment of 3 μ_B while the 8 outer Mn (marked as Mn_2 and Mn_3) are in a Mn^{3+} state and carry a localized moment of 4 μ_B coupled antiferromagnetically to the inner Mn. It is interesting to note that the bulk MnO which also corresponds to equal number of Mn and O atoms, has a NaCl structure and is an antiferromagnetic insulator. The D_2 structure of $Mn_{12}O_{12}$ can then be considered as a fragment of the bulk antiferromagnet network with uncompensated moments leading to a net magnetic moment.

While the above ionic picture can explain the overall magnetic moment, the existence of Mn^{4+} and Mn^{3+} ions would lead to highly localized moments at the Mn sites. This would lead to a weak intra-cluster magnetic coupling contrary to the observed behavior. Another evidence against the ionic picture comes from experiments on free $(MnO)_n$ clusters in beams. Ziemann and Castleman have generated $(MnO)_n$ clusters in molecular beams. Their observed mass spectra of $(MnO)_n$ shows magic sizes at n = 2, 3, 6, 9 and 12 MnO units. [20]. Note that these magic sizes are different from those observed for alkaline earth oxides i.e. $(MgO)_n$ or $(CaO)_n$ where the bonding is known to be ionic. This suggests that the bonding in $(MnO)_n$ clusters is more than simple ionic bonding. The beam experiments are also interesting from the point of deriving a structure. Based on the peaks at 3, 6, 9 and 12, Ziemann and Castleman suggest that the preferred structures of a free $Mn_{12}O_{12}$ cluster are stacks of the hexagonal $(MnO)_3$ rings. In particular, for $Mn_{12}O_{12}$ they propose a hexagonal tower shown in Fig. 6.2c. For the remainder of the paper we refer to this structure as the hexagonal-tower. Note that the tower structure is different from the D_2 structure proposed for a $Mn_{12}O_{12}$ clusters in Mn-acetate. This indicates that a free $Mn_{12}O_{12}$ may have very different properties than the corresponding clusters in the Mn-acetate and that the non-magnetic host in nanomagnets plays an important role in stabilizing the D_2 structure responsible for observed magnetic effect.

In this chapter, we review the geometry, electronic structure and the magnetic ordering in free and embedded $Mn_{12}O_{12}$ clusters. There are four basic issues. (1) The geometry and the magnetic ordering in free $Mn_{12}O_{12}$ clusters. In particular, if the hexagonal-tower is indeed the ground state configuration of a free cluster, what is its magnetic state? (2) The nature of electronic bonding and the magnetic coupling in ligated $Mn_{12}O_{12}$ clusters. (3) The role of non-magnetic host in stabilizing the D_2 magnetic structure. (4) The magnitude of the anisotropy energy as it controls the quantum tunneling. It is the energy barrier separating the various magnetic orientations. It is experimentally estimated to be around 50 K. The ab-initio calculation of the anisotropy energy is an extremely difficult task. In this work we briefly review a new scheme [11] to calculate the anisotropy and provide estimates of this quantity in the $Mn_{12}O_{12}$-acetate nanomagnets.

In Sect. 6.5 we briefly review the electronic structure scheme and Sect. 6.6 gives details of the theoretical calculations and results on smaller free Mn_nO_n fragments of the $Mn_{12}O_{12}$ cluster. Sect. 6.7 presents results on a free $Mn_{12}O_{12}$ clusters and Sect. 6.8 contains our studies on passivated $Mn_{12}O_{12}$ molecule. Finally Sect. 6.9 is devoted to a final conclusions.

6.5 Details of Theoretical Studies

This section reviews details of the theoretical studies on bare and embedded $Mn_{12}O_{12}$ clusters. For an embedded cluster we considered a core cluster surrounded by a layer of passivators. To make the electronic structure calculations computationally feasible the methyl terminators of carboxyl groups were replaced by H atoms. This substitution is not expected to affect the electronic and magnetic behavior of $Mn_{12}O_{12}$ clusters. Even with this simplification, the resulting cluster had 100 atoms and is shown in Fig. 6.2d. Note that the local environment of the Mn and O sites in the central $Mn_{12}O_{12}$ clusters is same as in the bulk acetate. This is needed to accurately reproduce the magnetic behavior. In addition, the calculated bond lengths of the passivated cluster were within 2% of experiment indicating that it provides a realistic description of the geometry, electronic and magnetic structure of the $Mn_{12}O_{12}$ cluster in the actual acetate.

The electronic structure studies were carried out using a linear combination of atomic orbitals molecular orbital approach. The particular set of computer codes we have used were developed by Pederson and Jackson and constitute the NRLMOL code [10]. The basis sets used to construct the molecular orbitals consist of gaussian functions centered at the atomic sites. The exchange correlation effects were incorporated via the recent functional proposed by Perdew et al. [21]. For details of the gaussian basis sets and the methodology used to solve the Kohn-Sham equations, the reader is refered to the original articles.

6.6 Geometry and Electronic Structure of Isolated $Mn_{12}O_{12}$ Clusters

As pointed out, the $Mn_{12}O_{12}$ clusters in the Mn_{12}-acetate have a D_2 symmetry and the Mn and O atoms are connected to the acetate and water molecules. To investigate the effect of the surrounding ligands on the geometrical arrangement and to examine if an isolated $Mn_{12}O_{12}$ cluster would have the same magnetic moment as the embedded cluster, we relaxed the core of the Mn_{12}-acetate. Figure 6.2a shows the unrelaxed geometry of this bare cluster. Upon relaxation under the D_2 symmetry constraint, it changes to the structure shown in Fig. 6.2b. The relaxed structure has a magnetic moment of 20.0 μ_B and in Fig. 6.2b we have shown the ordering of the various spins and the local moment at the various sites in the relaxed structure. Note that the bare cluster undergoes a significant relaxation. A comparison of the bond lengths shows that Mn_1–Mn_3 bond lengths in the bare and relaxed structure differ by 0.88 Å. This is largely due to the fact that the Mn_3 sites in the real structure have a coordination of six compared to only three in the bare cluster.

Figure 6.2c shows the tower structure proposed by Castleman and coworkers. The structure was optimised within a C_{3v} symmetry. This structure is 5.14 eV more stable than the relaxed structure shown in Fig. 6.2b. The cluster has no net magnetic moment. However, the individual Mn atoms have local moments of 5.0 μ_B, but the moments in different layers are antiferromagnetically aligned as shown in the figure. This shows that, although the acetate and the carboxyl radicals in the Mn_{12}-acetate are non-magnetic, they play a critical role in stabilizing the D_2 structure of $Mn_{12}O_{12}$ which makes the overall cluster magnetic. We would like to add that although the hexagonal tower is more stable than the bare D_2 structure and more stable than many other $Mn_{12}O_{12}$ clusters, it is never possible to guarantee that such a structure represents an absolute minimum. To further ascertain if the hexagonal tower is the most stable structure we have calculated the vibrational spectra for this molecule using the methods discussed in [22].

6.6.1 Vibrational Frequencies of the Hexagonal Tower

In order to establish that the hexagonal-tower is the ground state of a free $Mn_{12}O_{12}$ cluster, one needs to demonstrate that (1) it has the highest binding energy and (2) it is dynamically stable. This means that all vibrational frequencies of the cluster have to be real. To prove this point we have calculated the complete vibrational spectra of the hexagonal tower and the resulting vibrational density of states is displayed in Fig. 6.3. Note that all frequencies are real with the smallest frequency at $59\,cm^{-1}$. This shows that the hexagonal-tower is indeed dynamically stable and is therefore at least a metastable state.

We would like to add that the above condition is necessary but not sufficient to establish that the hexagonal tower is the ground state. To further proceed along these lines, we performed several more calculations in search for other possible ground-state structures. Starting from a 4-layered rock-salt structure for $Mn_{12}O_{12}$ with several different antiferromagnetic orderings of the Mn spins we performed geometry optimizations with a conjugate-gradient algorithm. In the first structure, the Mn atoms in the top and bottom layer are antiferromagnetically coupled to the the Mn atoms located at the inner two layers. This cluster had no net moment and converged to a geometry with the largest nuclear gradient of 0.0007 a.u. It looks quite similar to the hexagonal-tower but is found to be 0.31 eV higher in energy than that one. We believe that this structure is probably equivalent to the hexagonal tower and that further relaxation would allow this structure to converge to the hexagonal tower in Fig. 6.2c. In a second case we started from the same rock-salt $Mn_{12}O_{12}$ cluster but coupled the spins antiferromagnetically in alternating layers. This structure also had no net moment. After the geometry optimization the cluster retained the rock-salt geometry with the largest nuclear gradient of 0.0009 a.u. However, this structure was found to be 0.385 eV higher in energy than the hexagonal tower. In all these calculations we reduced the symmetry constraints significantly compared to the high symmetry hexagonal tower.

In addition to the structures presented in Fig. 6.2 we looked into other $Mn_{12}O_{12}$ arrangements with nonvanishing spin moments. None of these $Mn_{12}O_{12}$ clusters was found to be to be more stable than the hexagonal

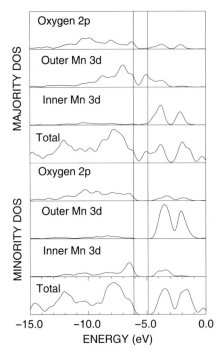

Fig. 6.3. Vibrational density of states for the hexagonal tower. As shown in Fig. 6.2c, this structure consists of a tower of four Mn_3O_3 hexagons. The net spin of this structure is zero. The inner two layers are ferromagnetically coupled. The outer two layers are antiferromagnetically coupled to the inner layers. The local moments on the inner layers are nominally $-5\,\mu_B$ while the local moments on the two outer layers are nominally $+5\,\mu_B$

tower. While other geometrical and magnetic configurations exist, the results of these calculations support the premise that the hexagonal tower is the best candidate for the ground state. However confirmation of this can only come by comparing our calculated vibrational modes to the experimentally determined vibrational spectra.

6.7 Electronic Structure of $Mn_{12}O_{12}$-Acetate

As discussed before, the mangenese-acetate contains $Mn_{12}O_{12}$ clusters surrounded by acetate radicals ($OOCCH_3$) and the water molecules. To render the theoretical calculations tractable, the methyl terminator on the acetate radicals were replaced by hydrogen atoms. With this replacement, the basic cluster containing the $Mn_{12}O_{12}$ core, the methyl terminators and the water molecules had 100 atoms. The starting geometry was derived from the experimental atomic positions of the Mn_{12}-acetate. The geometry was then optimized by calculating the forces on each atom and moving the atoms in

Table 6.3. Bond lengths found for several of the free D_{2d} $Mn_{12}O_{12}$ core (Fig. 6.2b) and for the ligandated molecule (Fig. 6.2d). Also included are bond lengths from the experimental X-ray data. All distances are in Angstroms. The labels on each atom correspond to those shown in Fig. 6.2

Bondlength	Isolated GND	Passivated	Expt.
Mn_1–O_1 (A)	2.09	1.91	1.90
Mn_1–O_1 (B)	2.14	1.93	1.92
Mn_1–O_2	2.28	1.92	1.88
Mn_1–Mn_1 (A)	3.14	2.90	2.94
Mn_1–Mn_1 (B)	2.82	2.85	2.82
Mn_1–Mn_3	2.60	3.48	3.45
Mn_1–Mn_2	2.65	2.82	2.77
Mn_2–O_2	1.84	1.90	1.89
Mn_3–O_2	1.90	1.90	1.89

the direction of forces using the conjugate gradient algorithm. It was found that the initial forces for the starting geometry were quite small and that the cluster only underwent minor relaxation.

Figure 6.2d shows the geometry of the relaxed cluster and Table 6.3 gives the various bond lengths in the bare and embedded $Mn_{12}O_{12}$ clusters and compares them with the experimental bond lengths. Note that the calculated bond lengths in the passivated cluster differ from the experimental bond lengths by less than 2%. Some of the bond lengths in the isolated cluster, on the other hand, differ by as much as 30%. This indicates that the non-magnetic host plays a critical role in stabilizing the approximate D_2 structure of the $Mn_{12}O_{12}$ cluster. Figure 6.4 shows the spin projected local density of states at the inner and outer Mn sites. Note that, whereas the majority spin states are mostly localized at the outer Mn sites, the minority spin states are mostly localized at the inner Mn. Further, the cluster has an overall magnetic moment of 20 μ_B. It is, however, interesting to estimate the local magnetic moments associated with various sites. For this, we calculated the total magnetic moment by integrating the total electron density within spheres centered around the sites. Our studies indicated that a sphere of around 2.5 bohr is able to capture most of the local moment. With such a choice, the studies give a moment of $-2.57\,\mu_B$ around the central Mn and 3.58–$3.63\,\mu_B$ around the outer Mn sites. Keeping in mind that the choice of the size of sphere is somewhat arbitrary and that the value of the moment depends on the size of the sphere, the above estimated values are close to the assignment of -3.0 and $4.0\,\mu_B$ based on the simple ionic model. One of the interesting features of the density of states is that the majority spin has a small gap and the minority spin has a large gap. The minority spin HOMO is 1.2 eV lower than the majority spin HOMO and the minority spin LUMO is 0.45 eV above the majority spin LUMO. This situation is similar to what is found in half-metallic ferromagnets.

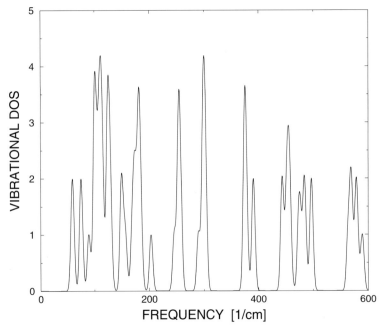

Fig. 6.4. Minority and Majority spin density of states for the $Mn_{12}O_{12}(RCOO)_{16}(H_2O)_4$ molecule. The majority electrons reside primarily on the eight outer Mn atoms and are primarily Mn 3d states. However the majority electrons also have a significant O 2p contribution indicating the the Mn – O bonding is partially ionic and partially covalent. Similarly, the minority electrons are primarily associated with the inner Mn 3d states but they also visit the oxygen atoms. The majority HOMO/LUMO gap is only 0.45 eV while the minority HOMO/LUMO gap is 2.08 eV. As such the electronic structure and magnetic ordering of this molecule can be characterized as approaching a half-metallic ferrimagnet

6.8 Magnetic Anisotropy Energy

The $Mn_{12}O_{12}$-acetate is found to be marked by a uniaxial anisotropy and there has been numerous studies using electron paramagnetic resonance [19] and other techniques to investigate the origin and the magnitude of the anisotropy. These studies indicate that the second and fourth order anisotropy terms are present. An approximate spin hamiltonian can be written as [18]

$$\Delta = \frac{B}{c}\langle S_z\rangle + A\langle S_z\rangle^2 + C\langle S_z\rangle^4 \ . \tag{6.6}$$

Here B is the external magnetic field, S_z is the z component of the spin, and A and C are constants indicating the strength of the second and fourth order terms. The best experimental estimates are $A/k_B = 0.556$ K and $C/k_B = 0.0011$ K.

6 Magnetism in Free Clusters and in $Mn_{12}O_{12}$-Acetate Nanomagnets

An ab-initio calculation of the anisotropy energy is a difficult task since it involves inclusion of spin orbit coupling and other relativistic effects. For uniaxial systems, the magnetic anisotropy is primarily due to spin-orbit coupling and an exact calculation requires a determination of the self consistent electronic structure including the spin orbit coupling and the calculation of the anisotropy energy using the resulting electronic states. In finite systems like clusters and nanoscale materials, the energy gap between the occupied and unoccupied manifolds is large and the inclusion of spin orbit term in the electronic structure calculation does not rearrange the filled and unfilled manifolds. In these cases, the anisotropy energy contribution due to spin orbit coupling can be easily calculated by first carrying out self consistent density functional calculation without the spin orbit coupling and the using perturbation theory to calculate the total spin-orbit energy as a function of the direction of the quantization axis.

In an earlier paper [11], two of the present authors had proposed a new approach to calculate the anisotropy barrier. The approach suggested a new way to calculate matrix elements of the spin-orbit operator. Instead of using the standard **L.S** representation for the spin orbit coupling, one starts from the basic expression

$$U(\mathbf{r}, \mathbf{p}, \mathbf{S}) = -\frac{1}{2c^2} \mathbf{S} \cdot \mathbf{p} \times \nabla \Phi(\mathbf{r}) , \qquad (6.7)$$

where **S** is the spin, **p** is the momentum operator and Φ is the coulomb potential. Expanding the single-electron wavefunction in terms of basis sets i.e.

$$\psi_{is}(\mathbf{r}) = \Sigma_{j\sigma} C_{j\sigma}^{is} f_j(\mathbf{r}) \chi_\sigma , \qquad (6.8)$$

where $f_j(\mathbf{r})$ are the spatial basis functions, the χ_σ are spinors ($\sigma = 1$ or 2) and the $C_{j\sigma}^{is}$ are the expansion coefficients of the Kohn-Sham orbitals. Hence, it is clear that the matrix elements of the spin orbit term require a knowledge of the Hartree potential and gradient of the basis function. As was earlier shown, matrix elements of the *coulomb potential* and the electronic spectrum are all that is needed to calculate the second-order change in the total energy of the molecule as one rotates the quantization axis. For details, the reader is refered to our earlier papers.

Using the above approach in conjunction with calculated electronic structure we calculated the anisotropy parameter A in 6.6. The actual calculation requires summation of the matrix elements connecting the occupied and the unoccupied states. In Table 6.4, we show the calculated values of the barrier for different cutoff energies and compare them with the experimental value of A. For all the cutoff's considered here, the calculated values are close to the experimental value. In fact, the inclusion of states beyond the valence states has practically no effect on the anisotropy constant. This shows that the anisotropy is primarily due to valence states. One can further analyse the

Table 6.4. Magnetic anisotropy energy (A/k_B) as a function of cutoff energy E_{cut} (above ϵ_F) for truncation of unoccupied state summation (6.7). Also included is the number of occupied and unoccupied states used in (6.7). The experimental data is from Ref. [18]

E_{cut} (eV)	N_{occ}	N_{unocc}	A/k_B (K)
6.8	804	381	0.558
13.6	804	730	0.557
27.2	804	1258	0.557
13.6 (valence only)	460	730	0.557
Experiment			0.556

theoretical contribution to anisotropy due to majority or minority states. In Table 6.5. we show the contribution of the various matrix elements to the overall second order anisotropy barrier. It shows that the barrier is primarily determined by the the overlap between the occupied and unoccupied states of different spin.

Table 6.5. Percent contribution to the anisotropy energy as a function of pairing between occupied and unoccupied states of different spins. Despite the fact that the smallest HOMO/LUMO gap is in the Majority–Majority channel, these pairs contribute only 13% to the total anisotropy energy. It is the antiparallel pairs of occupied and unoccupied orbitals, residing in spatially similar parts of the molecules, that are primarily responsible for the formation of the anisotropy barrier

Spin Pairing	HOMO/LUMO Gap	Anisotropy Contribution
Minority–Minority	2.08 eV	1 %
Minority–Majority	1.63 eV	21 %
Majority–Minority	0.89 eV	65 %
Majority–Majority	0.45 eV	13 %

6.9 Conclusions and Extension to Fe$_8$ Nanomagnets

The present review article shows that while the reduction in size leads to discrete angular mometum states, the Stern-Gerlach experiments are not able to observe space quantization in free atomic clusters. Part of the problem is the reduction in the anisotropy energy that leads to super-paramagnetic relaxations. We then discussed that the space quantization can be observed in macroscopic magnetic hysteresis loops provided the magnetic clusters could satisfy certain conditions. $Mn_{12}O_{12}$-Acetate presents such an example. We showed that the non-magnetic hosts play a critical role in the observed magnetic behavior of $Mn_{12}O_{12}$-Acetate. For free $Mn_{12}O_{12}$ clusters, a tower obtained by stacking hexagonal Mn_6O_6 rings is more stable than the structure observed in the acetate and it has no net magnetic moment. The D_2 structure of $Mn_{12}O_{12}$ clusters in acetate is stabilized by the acetate and water molecules. Recent experimental work indicates that the magnetic relaxation may be assisted by phonons. This indicates that the host not only stabilizes a structure which makes it magnetic, but also plays a critical role in the magnetic relaxation. The system also provides grounds to test the new approach to magnetic anisotropy. The results show that it not only provides quantitatively accurate values of the anisotropy energy but also offers a framework to understand the physical origin of the anisotropy. We would like to add that we have recently extended our studies to Fe_8 class of nano-magnets. Our studies provide good agreement with experiment on the directions of the easy, medium and hard axis. Also, the oscillations in the tunnel splitting determined from our anisotropy parameters match the experimentally observed oscillations.

Acknowledgements. SNK was supported in part by the Department of Energy (DE-FG02-96ER45579). MRP was supported in part by the ONR Molecular Design Institute (N00014-98WX20709).

References

1. R.S. Berry, J. Burdett and A.W. Castlemann, Jr.; Eds., Small Particles and Inorganic Clusters, Z. Physik, **26** (1993).
2. R.H. Kodema, Jour. Mag. Mag. Mat. **200**, 359 (1999).
3. F. Liu, M.R. Press, S.N. Khanna and P. Jena, Phys. Rev. B **39**, 6914 (1989).
4. W.A. de Heer, P. Milani and A. Chatelain, Phys. Rev. Lett. **488** (1990).
5. I.M.L. Billas, A. Chatelain and W.A. deHeer, Science **265**, 1682 (1994) and references therein.
6. A.J. Cox, J.G. Louderback and L.A. Bloomfield, Phys. Rev. Lett. **71**, 923 1993.
7. S.N. Khanna and S. Linderoth, Phys. Rev. Lett. **67**, 742 (1991).
8. J.Q. Xia et al., Phys. Rev. Lett. **68**, 3742 (1992).
9. P. Jensen and K.H. Bennemann (Private Communication).
10. M.R. Pederson and K.A. Jackson, Phys. Rev. B, **41**, 7453 (1990).
11. M.R. Pederson and S.N. Khanna, Phys. Rev. B **60**, 9566 (1999); K.A. Jackson and M.R. Pederson, Phys. Rev. B **42**, 3276 (1990).
12. R. Fournier and D.R. Salahub, Surface Science **238**, 330 (1990).
13. J.R. Friedman, M.P. Sarachik, J. Tejada and R. Ziolo, Phys. Rev. Lett. **76**, 3830 (1996).
14. A. Barr, P. Debrunner, D. Getteschi, C. Schulz and R. Sessoli, Europhys. Lett., **35**, 133, (1996).
15. E.M. Chudnovsky and L. Gunther,Phys. Rev. Lett. **60**, 661 (1988).
16. L. Thomas, A. Caneschi and B. Barbara, Phys. Rev. Lett. **83**, 2398 (1999).

17. T. Lis, Acta Crystallogr. Soc. B **36**, 2042 (1980).
18. A. Fort, A. Rettori, J. Villain, D. Gatteschi and R. Sessoli, Phys. Rev. Lett. **80**, 612 (1998).
19. S. Hill, J. A.A.J. Parenboom, N.S. Dalal, T. Hathaway, T. Stalcup and J.S. Brooks, Phys. Rev. Lett. **80**, 2453 (1998).
20. P.J. Ziemann and A.W. Castleman Jr., Phys. Rev. B**46**, 13480 (1992).
21. J.P. Perdew, K. Burke and M. Ernzerhof, Phys. Rev. Lett. **77**, 3865 (1996).
22. D.V. Porezag and M.R. Pederson, Phys. Rev. B **54**, 7830 (1996).

7 Size Effects in Catalysis by Supported Metal Clusters

A.A. Kolmakov and D.W. Goodman

7.1 Introduction

Clusters are a transition state of matter between molecules and the bulk material in that their electronic and geometric properties change from molecular-like to bulk-like with increasing size. Since the chemical reactivity of a cluster is intimately related to its electronic structure, clusters have unique chemical properties, different from those of isolated molecules and the corresponding bulk material. Indeed, ensemble-specific variations in cluster reactivity and a general tendency to approach bulk-like properties with increasing cluster size were found in early cluster molecular beam studies [1–3]. Of fundamental interest for supported clusters is whether such clusters exhibit the reactivity of their gas-phase counterparts and to what extent does the support alter their chemical properties. This issue has been addressed in applied catalysts where highly dispersed metals are frequently employed to optimize reactivity by maximizing the exposed metal surface area. However, the majority of available data indicated [4–6] that neither the certain size range nor the size-dependence of reactivity can be explained simply by noting the degree of metal atom coordination. Therefore allocating these complicated properties to collective phenomena, e.g. electronic properties, the morphology, cluster-support interaction, etc., is essential for the development of a molecular-level understanding of structure/selectivity/activity relationships in heterogeneous catalysis.

Variations in the conductance, porosity, preparation, and activation of "real world" catalysts limit the application of most surface analytical methodologies. The morphological complexity of the "real world" systems also limits the application of tractable theoretical models. However, metal clusters deposited under clean UHV environments onto ordered thin oxide films or semiconducting single crystals are model catalysts with acceptable simplicity and reproducibility of use with an array of X-ray, electron, and ion spectroscopies and microscopies [7–9]. Using this "surface science" approach, significant advances have been made during the last decade in correlating the chemical reactivity of supported clusters with their electronic/geometrical structures and the nature of the support/cluster interaction.

A number of excellent reviews have appeared over the past several years highlighting the various approaches to the study of supported metal clusters [10–17]. These reviews cover: (i) the different preparation and characterization strategies for model catalysts; (ii) advanced experimental methodologies; (iii) the adsorption and catalytic properties of supported clusters; and (iv) the cluster-support interaction.

Due to the extensive literature on this subject, this chapter only briefly summarizes the key issues relevant to structure-selectivity relationships in catalysis. Some examples of recent studies where size effects of supported clusters are manifested as marked variations in chemical reactivity are described. In this context, recent results in the preparation and characterization of nanosized model catalysts are also presented.

7.2 Methodology

7.2.1 Thin Oxide Films as a Model Support

Achieving an atomic-level understanding of catalytic processes that take place on nanoclusters has been hampered by the inherent complexity of commercial catalysts, typically operated at elevated temperatures and pressures. Several experimental strategies have been used to effectively address these complex industrial systems. One of the widely used approaches is substitution of the porous, non-conducting, oxide support by oxide single crystals or well-ordered thin films [7,8,17]. Such films can be prepared in ultrahigh vacuum (UHV) with high quality, high purity, and careful control of the stoichiometry. With a thickness of 0.5–3.0 nm, these ordered oxide films exhibit sufficient thermal and electrical conductivity to employ a variety of surface spectroscopic and microscopic techniques over a wide temperature range. Considerable work has been carried out by several groups addressing the preparation and characterization of ultra-thin oxide films (see reviews in [7,8,18] and references therein). The similarity of the basic chemical and physical properties of these oxide films to the corresponding high surface area analogs has been demonstrated [18–20].

Due to the typically large oxide-support lattice mismatch at the interface (Al_2O_3/NiAl (100)) [23,24], thin film preparation via oxidization of the native material typically results in disordered surfaces. However, good quality metal oxide films can be obtained with this method using a suitable alloy as a substrate (Al_2O_3/NiAl (100)) [23,24]. These reproducible, high quality films with a well-characterized thickness are currently in use by a number of groups as a support for model catalysts. A more flexible approach, suitable for the production of a variety of films with varying thicknesses and stochiometries, is the growth of epitaxial oxides via evaporation of the corresponding metal in ca. 10^{-6} torr oxygen ambient onto a single crystal support [14,15,25]. The film-support lattice mismatch can be minimized by selecting a suitable substrate material. This method can produce high quality films; however, doing so requires the optimization of numerous film growth parameters.

7.2.2 Cluster Deposition: Density, Size and Control of Morphology

Vapor Deposition. Vapor deposition of metals onto oxide surfaces is the simplest and most widely used method for the synthesis of supported clusters [10–12], [15,26]. Several elementary steps, i.e. adsorption, diffusion, nucleation, cluster growth, etc., are involved in cluster formation which vary with the metal, the support oxide, and the experimental conditions. However, these variables provide wide flexibility in controlling cluster density, size and morphology. For homogeneous nucleation, the cluster density can be enhanced by increasing the deposition rate and/or decreasing the adatom mobility. For the same equivalent coverage, an increase in the deposition rate and/or a decrease in the adatom mobility lead to a decrease in cluster size.

An alternative method of decreasing cluster size is the creation of artificial nucleation centers (heterogeneous nucleation) on the support by the controlled formation of defects [27] or the pre-adsorption of a reactive adsorbate before cluster deposition [28]. As the saturation density of the cluster is approached, subsequent metal deposition leads only to metal cluster size growth.

The growth mode and morphology of nanoclusters created near thermodynamic equilibrium conditions depend upon the interplay of the surface free energies of the metal, the oxide, and the metal-oxide interface. For most of the catalytically important metals, wetting is thermodynamically unfavorable and the deposited metal forms three dimensional (3D) aggregates (see Fig. 7.1). In this case the shape of the aggregates corresponds to a minimum in the surface/interface energy and can be described in terms of truncated Wulff ployhedra [29,30]. However, in attempts to synthesize extremely small clusters, i.e., < 2 nm, the growth conditions are often quite far from thermodynamic equilibrium and kinetic limitations dominate the cluster morphology. Under such conditions the cluster can grow two-dimensionally to a critical coverage, after which a pseudo-layer-by-layer mode can dominate [31–33]. This behavior is particularly important for supported catalysts since the 2D structure itself and the subsequent 2D–3D thickening transition might contribute to strong size dependence in the reactivity of small clusters and their stability as well. (See discussion below)

One general disadvantage of the vapor deposition method is that a relatively broad cluster size distribution is a consequence due to the statistical nature of the nucleation process. The cluster distribution becomes critical when chemical activity shows a marked cluster size dependence, i.e., a narrow cluster size range exhibits special catalytic activity. Recently this size distribution limitation has been circumvented by size selective cluster evaporation using laser oblation [34,35]. The metal clusters so formed have size-dependent surface-plasmon absorption resonances that can lead to cluster desorption upon excitation [36]. Therefore, by turning the laser light to one particular portion of the cluster size distribution, the total cluster size distribution can

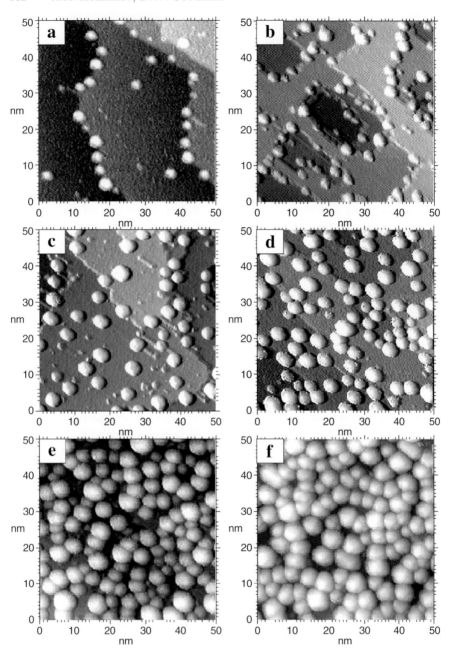

Fig. 7.1. The change in morphology of the TiO_2 surface upon Au deposition. The STM data conditions: CCT, 2V, 2nA. (**a**) 0.1 ML, the extended defects are the major nucleation centers for the clusters growth: (**b**) 0.25 ML, the point defects become populated with clusters; (**c**) 0.5 ML, the diameter and height of the clusters increase; (**d**) 1 ML, along with cluster sizes, the density of clusters increases due to continuing growth of small intermediate clusters; (**e**) 2ML; and (**f**) 4 ML, the clusters are in contact with each other and coalescence begins

be significantly sharpened by desorption of the portion excited. By varying the excitation energies, a cluster size standard deviation of 0.13 was achieved for Ag clusters with an average diameter of 10 nm [35].

Ligand-Stabilized Clusters. Metal clusters can be nucleated in solutions and synthesized with a given size by interrupting nucleation with surfactants. Recent successes in the chemical synthesis of these organometallic compounds have yielded nearly monodispersed metal clusters having from 1 to 8 shells of face-centered, close-packed (fcc) atoms [37]. Even the production of hollow [38] or multishell onion-like bimetal particles is now possible [39]. These clusters exhibit exciting chemical and quantum-sized properties such as Coulomb blockades in supported clusters [40] and single electron inter-cluster transport [41]. The great potential of this approach for the synthesis of catalysts and nano-electronic devices has sparked considerable research during the last decade [42]. The deposition of metal nanoclusters on planar oxide supports can be carried out by wet impregnation or by evaporation in UHV as long as the thermal stability of the organometallic precursor is sufficiently high [43]. The activation of the catalyst requires the removal of the organic ligands from the cluster. This ligand removal and subsequent reactivity measurements are the subject of ongoing studies.

Size Selected Clusters. The ideal choice for a planar, oxide-supported model catalyst is one with dense, mono-dispersed metal nanoclusters and one with well-defined and controllable cluster size. Such models have been developed over the last 15 years in several laboratories using mass-selected cluster deposition coupled with molecular beams [44–46]. Many experimental challenges (e.g., the creation of intense cluster beams with a narrow energy spread, mass selection, cluster soft landing, and cluster stabilization on the surface) have been resolved using this methodology. Currently, for a number of metals, size-selected cluster beam sources are available with kinetic energies less than 1 eV per atom with cluster fluxes sufficient to achieve a significant fraction of a monolayer of clusters in an acceptable time period. The surface mobility of the clusters is generally inhibited at low support temperatures or by pre-formed point defects. An additional advantage of this approach is the possibility for directly comparing the chemical reactivity of supported clusters with their gas phase counterpart and in so doing, shed light on the nature of the cluster-support interaction. Since the deposited clusters are pre-formed in the beam, their morphology is typically quite different from clusters grown directly on the surface. Thus, care has to be taken when comparing the reactivity of clusters produced with these different approaches. The current limitation of the method is the inaccessibility of clusters larger than $N > 100$. However, this limitation should be circumvented in the near future with further development in mass filters and cluster sources.

Nano-Lithography. This complementary approach for the engineering of model catalysts has been recently developed [47–51]. The approach is the synthesis of planar catalysts using modern nano-lithographic methods. In principle, nanometer-sized clusters with variable diameters and compositions, either amorphous or crystalline, can be synthesized using this approach. In addition pre-patterning of the active support can be carried out prior to cluster deposition. Because catalysts synthesized using this method can be patterned, i.e., well-defined cluster arrays, this approach is especially suitable for the study of kinetic phenomena at the surface such as lateral mass transport, cluster sintering, adsorbate spill-over, reactant re-supply via the support, etc. An example of the thermal and chemical stability that can be achieved at present is shown in Fig. 7.2 [48]. With current state-of-the art methodologies, arrays with well-fined separations of metal clusters in the range of 10–100 nm have been prepared on the surface of metal or oxide single crystal or films [50,51]. In comparison with other cluster deposition methods, nano-lithography requires multiple processing steps and therefore is relatively time consuming.

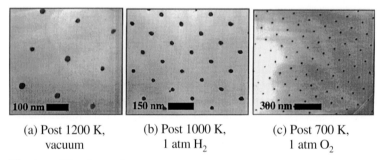

(a) Post 1200 K, vacuum

(b) Post 1000 K, 1 atm H_2

(c) Post 700 K, 1 atm O_2

Fig. 7.2. The thermal and chemical stability of ~ 20 nm Pt nanoclusters on silica fabricated using electron beam lithography. TEM images showing that the array of Pt nanoclusters remained intact during heating in: (**a**) vacuum at 1200 K; (**b**) 760 Torr of H_2 at 1000 K; and (**c**) 760 Torr of O_2 at 700 K (from [48])

7.2.3 Analytical Tools: Spectroscopy and Microscopy

Traditional Laboratory Techniques. Although valuable information concerning adsorbate bonding parameters, adsorption/desorption, and reaction kinetics can be obtained using bulk, non-conducting oxide samples and molecular beams, photo- and electron- stimulated spectroscopies, ultra-thin oxide films are responsible for much of the recent progress toward an understanding of the electronic properties of nanoclusters and the nature of their interaction with the support. In films 0.5–3.0 nm thick, the tunneling current through the film is sufficiently large to neutralize the surface charge induced by the impinging electrons, ions or energetic photons. Thus, clusters supported on oxide films and the oxide films themselves, become tractable for a range of surface science techniques including X-ray photoelectron spectroscopy (XPS), Auger spectroscopy (AES), high resolution electron energy loss spectroscopy (HREELS), electron/ion scattering and scanning probe/electron microscopies. Comprehensive reviews demonstrating the effectiveness of these methods for studying of the evolution of electronic structure and particle morphology with the cluster size have appeared recently [52–55]. Particular examples will be discussed in the following two sections, highlighting recent trends and developments in experimental tools relevant to the characterization of supported nanocluster and their chemical reactivity.

Synchrotron Radiation (SR). The major limitation of most laboratory spectroscopic techniques is the lack of spatial and energetic specificity. For example, in routine XPS measurements the sample is typically illuminated with a narrow, yet non-monochromatic, X-ray line of 1–10 KeV, and photoelectrons acquired form a spot of ca. 0.1–1 mm diameter. Since deposited nanoclusters usually cover only a small fraction of a planar support, XPS features of the clusters are generally weak and superimposed on the support background signal. With the availability of synchrotron radiation (especially the third generation machines), several advantages have been realized: (i) the intensity of the synchrotron light at any particular wavelength in the VUV and X-ray regions exceeds laboratory sources by several orders of magnitude, allowing fast and high-resolution measurements to be made even for dilute species such as supported nanoclusters. In addition, the ultimate spectral resolution is not dependent on the light source but on the monochromator and detector. A resolution of $E/\Delta E \sim 10^4$ can be achieved within the soft X-ray range [56] which, in conjunction with modern electron analyzers, permits one to probe the details responsible for the broadening, shifting, and splitting of absorption and photoemission lines; (ii) fine-tuning of the excitation photon energy is possible to significantly enhance the sensitivity allowing one to probe a particular element or even a particular chemical bond or interface using photon energies resonance with a particular component of a specimen; and, (iii) powerful spectroscopies like near edge X-ray absorption fine structure (NEXAFS) and extended X-ray absorption fine structure (EXAFS) are available for analysis of electronic and geometrical structures of model catalysis. NEXAFS relies on the fine structure of the electronic transitions within ~ 10 eV of the core level absorption edge which is characteristic of a particular exited atom or molecular bond [57]. EXAFS probes structural information via the relatively weak modulations in the absorption cross-section due to an interference with the wave functions of outgoing electrons and electrons back-scattered from the nearest coordination shells [58]. The relevance of these two methods to nanoclusters and their reactivity is two fold: (a) both probe element-specific, deep core levels with information directly related to the local electronic and geometrical structure can be obtained at the nanoscale level; and (b) these are among the few techniques sensitive not only to adsorbates but are also suitable for probing catalysis in situ under realistic reaction conditions. Based on the relationship between gas phase and solid state absorption cross-sections and using various detection methods, e.g., secondary or Auger electrons and soft X-ray fluorescence, a variety of elegant and powerful experimental methods for acquiring such data have been developed, see recent reviews [59,60].

Since SR is highly polarized, virtually all the methods mentioned above are relevant to the magnetic properties of nanoclusers. Even greater opportunities in this respect for the study of nano-clusters will be made available with more extensive development and utilization of free electron lasers [61].

Spectromicroscopy. The intense interest in the chemical, structural and electronic properties of heterogeneous surfaces and the availability of very bright SR sources has led to the rapid development of spectro-microscopies with spatial resolution at the nano-scale level. These powerful techniques have already made a significant impact on the understanding of model catalysis [62,63] and will play an increasingly larger role in the near future in advancing our understanding in this area. Generally these measurements can be divided into two categories: (i) scanning microscopes; and (ii) photoelectron emission (PEEM) and low energy electron microscopies (LEEM).

In scanning microscopies, soft X-rays are focused into a tiny area by a Fresnel zone plate or Schwartshild objectives and the focused beam is scanned over the surface [64,65]. "Chemical map" images are acquired while setting the energies of the primary monochromator and the electron analyzer to excite and detect, respectively, the appropriate core level electrons. In addition complete XPS, NEXAFS, and EXAFS can be acquired from a particular point of interest, thus providing compositional, chemical and geometrical structure at the nano-scale. Currently the maximum lateral resolution achieved for these kinds of microscopes is ~ 100 nm, within the size range of some oxide-supported metal catalysts. The lateral resolution is determined by the precision of currently manufactured zone plates, and remains well in excess of the diffraction limit of ca. 2 nm for soft X-rays ($500-10^3$ eV).

PEEM and LEEM generates magnified images using electrons emitted exclusively from a particular small spot on the surface detected with an electronic lens while irradiating the sample with photons or slow electrons, respectively [66]. The introduction of an energy filter in the imaging optics allows imaging with energy selected electrons. Scanning the photon energy with the primary monochromator allows images sensitive to the absorption edges to be acquired, thus X-ray absorption near edge spectroscopy (XANES) or NEXAFS can be performed with excellent spatial resolution. A thorough review of these imaging and spectroscopic microscopes can be found in [67]. It is noteworthy that enhanced resolution (1–2 nm) is anticipated for the newly constructed instruments at BEESY II and ALS [68]; therefore methods should become available in the near future for nano-cluster imaging and local spectroscopy in real time.

Scanning Probe (SPM) and Transmission Electron Microscopies (TEM): In Situ Imaging Under Reaction Conditions. Scanning probe microscopies provide local information regarding the details of the surface structure. In general, scanning tunneling microscopy (STM) and spectroscopy (STS) have been used extensively in the past on planar metal catalysts and semiconductors. The most recent advancements, including the imaging and chemical identification of adsorbates with STM, have been recently surveyed [69,70]. The imaging of molecular diffusion and specific reaction steps in real time on planar surfaces is now possible in addition to carrying out inelastic tunneling spectroscopy of a single, adsorbed molecule [71]. Potentially, all these spectroscopic techniques are applicable to supported nano-clusters as well. The problems associated with the imaging of poorly conducting materials such as oxides and thin films have been addressed in [72,73]. The imaging of metal nano-clusters supported on low conducting materials is, in principle, a challenging task in that the imaging can be accompanied with artifacts not necessarily related to tip quality. Difficulties arise due to tip-induced band banding and the formation of nanoscale Schottky barriers at the cluster-support interface [74]. However, as was verified in combined studies on a single interface using STM and scanning electron microscopy (SEM),

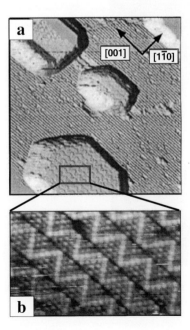

Fig. 7.3. (a) a STM image (50 × 350 nm) of an encapsulated Pt cluster on $TiO_2(110)$ after a high temperature treatment; and (b) an atomically resolved image of an encapsulated cluster (from [78])

reliable (within ~20% accuracy) cluster size measurements can routinely be obtained on oxide supports using well-defined tips and optimization of the tunneling conditions [75]. Very recently, atomic resolution of flat clusters was achieved [76] and the encapsulation of Pt clusters on titania was directly imaged (Fig. 7.3) [77,78]. Efforts to apply scanning probe microscopies under

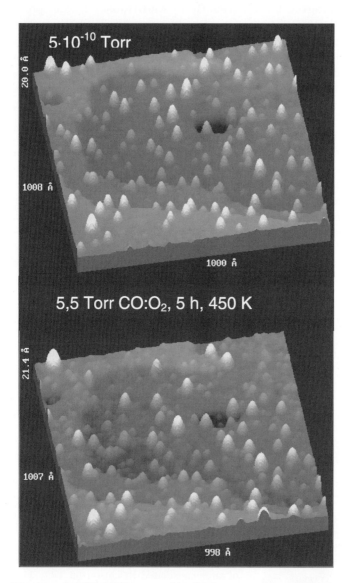

Fig. 7.4. A CCT STM image of Au/TiO$_2$ under CO oxidization reaction conditions, targeting the same area over eleven orders of magnitude of pressure

"real world" reaction conditions is in its infancy, but has revealed new information regarding the actual, working catalyst. Atomic resolution of a planar metal catalyst was achieved via in situ STM with a reactive gas pressure spanning ca. 14 orders of magnitudes (from UHV to atmospheric), bridging the so-called "pressure gap" [79]. Exiting new information regarding pressure-induced surface restructuring was obtained in these studies.

To explore the morphological evolution of supported clusters under catalytic reaction conditions, similar studies were undertaken in our laboratory on oxide supported noble metal catalysts [80]. In situ imaging of Au clusters on TiO_2 (110) showed the effects of sintering and surface morphology modifications while changing the pressure of the reactions ($CO:O_2$) from UHV to the actual working conditions of the catalyst (Fig. 7.4).

The recent insertion of a pressurized sample stage (microreactor) inside the column of a high resolution TEM (HRTEM) has transformed traditional electron microscopy into a powerful tool for studying supported model catalysts under reaction conditions. Atomic resolution can be achieved under elevated pressure conditions with simultaneously compositional (EDX, PEELS) and structural analysis (electron diffraction) in real time [81]. In Fig. 7.5 the in situ imaging of the encapsulation of a Pt nanocluster via a strong metal support interaction (SMSI) is demonstrated. An image of an active (300 °C in H_2) cluster and a second image subsequent to its deactivation (450 °C in H_2) are shown. In addition, the aggregation of small ~ 2 nm Pt clusters due to sintering is indicated by the highlighted arrow. The combination of the resolving power of modern microscopy with state-of-the-art, adsorbate-sensitive spectroscopies such as polarization modulation IRAS [82], frequency

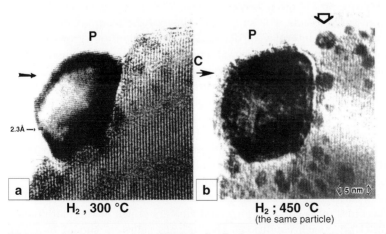

Fig. 7.5. An environmental-HREM of a finely dispersed Pt/titania catalyst under reaction conditions on the atomic scale: (**a**) after in situ activation in H_2 gas at 300 °C (Pt cluster lattice resolved); and (**b**) the same cluster (P) at 450 °C also in H_2 becoming covered with a Ti-rich sub-nm overlayer of C (from [81])

modulated laser Raman [83,84] and sum frequency generation [85] provide opportunities for the complete characterization of catalysts on an atomic level with realistic, operating reaction conditions.

7.3 Cluster Size and Reactivity

The reactivity of metal clusters in the gas phase has been extensively studied and the corresponding activity correlated with electronic structure and structural stability [1–3]. The specific size dependence in cluster reactivity for gas phase suggests similar trends may apply to supported clusters, at least those relatively small clusters that can be probed in the gas phase. However, deposition and growth of clusters on the surface of a planar oxide support can dramatically change the electronic structure, charge status, morphology and dimensionality of nanoclusters. Consequently, size effects in chemical reactivity for supported clusters very well may be decidedly different from those found for gas phase clusters because of support-cluster interactions. We review briefly only those structural and electronic factors that are moderately well understood. For consistency, turnover frequency (TOF) or the reaction rate per surface atom, will be used for catalytic activity. Because of their complex interrelationship, the complete separation of geometric and electronic effects is not possible; however, when appropriate, emphasis on the role of each will be made.

7.3.1 Geometric Factors

Low coordinated atoms often become active centers for chemical reaction due to unsaturated bonds. For supported clusters the shape of the cluster is determined by the interplay of deposition, nucleation and growth kinetics, and thermodynamics. With growth conditions far form thermodynamic equilibrium, the shape of the cluster depends on local kinetics and on the interaction of the support with the cluster. This shape can be a 2D epitaxial island or in many cases a 3D amorphous aggregate. In either case the surface atoms are generally equivalent and represent an ensemble of low coordinated metal centers. Obviously, the fraction of surface atoms increases with decreasing cluster diameter. The perimeter atoms of the cluster are often considered to be special active centers since their coordination includes the bonding of the cluster atoms to the atoms of the support [86]. The relative abundance of these interface atoms increases with decreasing cluster size. The situation becomes more complex for metal clusters grown at thermodynamic equilibrium. Under such conditions, theoretically, clusters should have the shape of truncated Wulff's polyhedra (see [5,6]) having inequivalent, low coordinated surface atoms at the edges, corners, and interiors of various facets (Figs. 7.3, 7.6). In reality, clusters are actually incomplete ployhedra and have incomplete facet layers leading to atoms with greatly varying

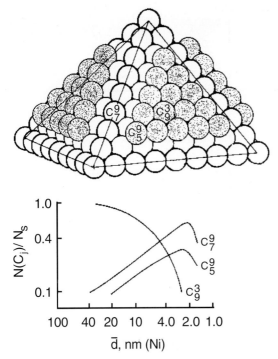

Fig. 7.6. A face-centered cubic (fcc) octahedron. The shaded atoms represent incomplete layers. In the lower part of the figure the fraction of exposed atoms, $N(C_j)/N_s$, for various atoms is plotted as a function of the mean cluster size, C_j is the coordination number and N_s is the number of surface atoms (adapted from [6])

coordination. Depending on the nature of active sites or facets mentioned above, the proportion of these active centers (so-called fraction exposed, FE) exhibits marked variations with cluster size for clusters smaller than ~ 5 nm (Fig. 7.6, bottom). This is precisely the cluster size range where the TOF changes dramatically in many structure sensitive reactions. It is noteworthy that the observed size variation of the reactivity in many cases matches closely the theoretical predications based on the fraction-exposed concept. However, obviously this approach oversimplifies the actual system due to the size and shape distribution of the actual metal clusters and the changes that occur under reaction conditions.

Another effect that could strongly influence the electronic structure of small clusters is the lattice contraction that accompanies a decrease in cluster size. This is also a size effect related to the increase in the fraction of surface atoms. As with liquid droplets, a decreasing cluster diameter increases the internal pressure within the cluster leading to a reduction of the lattice parameter a such that $\Delta a/a = -4/3 \cdot Q \cdot F/d$, where Q is the compressibility of the metal and F is the surface tension. For the supported clusters that have

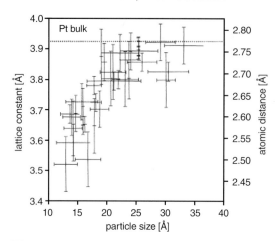

Fig. 7.7. The lattice constants and interatomic distances of Pt clusters grown on A_2O_3 NiAl(110) as a function of their size. The horizontal bars represent the width and the length of the particular clusters, respectively, and the vertical bars correspond to the error (from [87])

been studied, a change in the lattice constant of 5–10% has been observed (see Fig. 7.7) for clusters in the range of 1–3 nm, reaching the bulk value for clusters larger than ~ 4 nm [87].

An additional strain and lattice contraction/expansion can be induced even for 2D clusters at the early stages of epitaxial growth due to cluster-support lattice mismatch. Defect formation can be initiated inside the growing cluster to release this strain, an effect that has been discussed in the literature [88].

At room temperatures, in spite of the thermodynamically preferable 3D (Volmer-Weber) mode, deposition of many transition metals onto oxide surfaces leads to pseudo-layer-by-layer or 2D cluster growth [89,90]. After some critical 2D cluster size is reached, the growth mode switches to the 3D regime. The reason for this initial wetting behavior is the presence of a relatively large potential upstep barrier, E_1, for metal atoms to diffuse from the oxide surface to onto a 2-D metal island Fig. 7.8b. The temperature is assumed to be large enough to allow adatoms to diffuse on both interfaces and sufficient to overcome the downstep barrier, E_{-1}. As soon as the 2D island becomes large enough for 3D nucleation, a second layer begins to grow. The critical size is dependent on the particular metal and oxide involved as well as on the experimental parameters such as temperature, flux and defect density. The average critical size for most metals studied corresponds to a few up to a few hundred atoms. Since the stability of the metal-metal bond generally exceeds that of the metal-oxide bond, one can consider these 2D metal atoms as "coordinatively unsaturated" with respect to the 3D case. Thus, in analogy to the low coordinated surface or defect sites considered above, 2D clusters

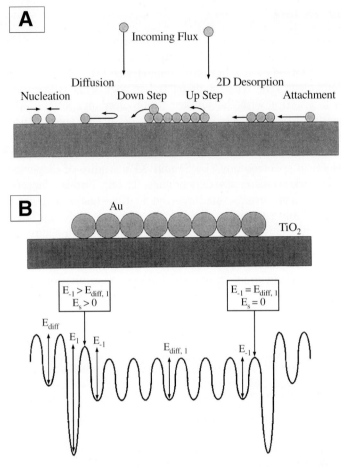

Fig. 7.8. (a) A schematic representation of various elementary atom migration steps involved in metal vapor deposition and island formation with a qualitative energy diagram (b) showing the potential energy encountered by a diffusing adatom on top of the surface. The various activation barriers are also shown on the diagram. A schematic representation of the surface is aligned along the vertical axis with the potential energy surface. The left-hand side of the island shows $E_{-1} > E_{dift,1}$, while the right-hand side of the island shows $E_{-1} = E_{dift,1}$ (from [89])

should demonstrate completely different reactivity properties compared to their 3D counterparts. Indeed, this has been shown to be the case and specific examples will be discussed in the following section. In addition, due to the marked size dependent 2D-3D transition, low dimensional clusters are metastable and are irreversibly converted into 3D structures upon annealing or exposure to an adsorbate. This may be one of the reasons for the decline in activity with reaction time of certain oxide-supported metal catalysts.

7.3.2 Electronic Factors

An excellent example of a correlation of electronic structure and chemical activity of clusters is shown in Fig. 7.9 where the relative reactivities of gas phase Fe clusters are plotted as a function of cluster size [91]. As discussed above, the size dependent reactivity of deposited nanoclusters is strongly influenced by their interaction with the support. Photoelectron spectroscopy, because of its sensitivity to the electronic structure of surface species, was among the first techniques used to explore size dependent electronic structure of supported metal clusters and related cluster-support effects. Indeed, early ultraviolet photoelectron spectroscopy (UPS) and XPS studies of dispersed nanoclusters showed marked cluster size dependency [92–96]. Two key factors, initial and final state contributions, are involved in the cluster ionization process, and both contribute to spectral shifts. In addition to initial state effects associated with free clusters, the interaction of clusters with a support contributes additional initial state terms due to chemical bond formation (oxidization) with interfacial atoms, cluster dimensionally, and the initial cluster charge. For supported clusters, finial-state contributions are a function of the overall screening, the Coulombic response to the formation of the photoelectron, and the induced polarization of the support (see Fig. 7.10). All of these factors are cluster size dependent. The major experimental task therefore is to determine: (i) whether the observed spectral changes are artifacts due to adsorbate-induced band bending or surface charging; and (ii) if intrinsic

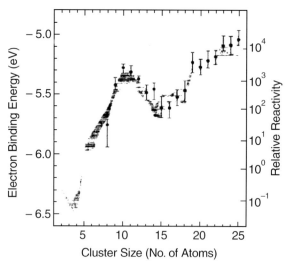

Fig. 7.9. A comparison of measured ionization thresholds (to the left) with the intrinsic reactivity of Fe clusters. The gray band reflects the uncertainty in the ionization threshold measurements, while the vertical lines indicate uncertainties in the reactivity results, acquired from measurements of Fe_x depletion by reaction with D_2 and H_2 (from [91])

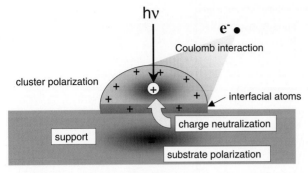

Fig. 7.10. The basic ionization and screening steps for supported clusters

electronic modifications actually occur, what is the nature of these changes. Assessing the relative contributions is complex and requires careful control of the experimental parameters.

Typical UPS spectra acquired as a function of cluster size are shown in Fig. 7.11 [96]. The results to date can be summarized as follows: (i) the valence band narrows and the band centroid shifts toward higher binding energies with decreasing cluster size; and (ii) the Fermi edge broadens and, along with the valence band features, move toward higher binding energies at the small cluster limit. Smaller clusters exhibit an increase in binding energy (positive core level shift) even for those metals that exhibit a negative surface-atom core level shift in the bulk.

Band narrowing is purported to be related to finite size effects, i.e., the reduction of the atom coordination number, N_c, in small clusters causes a reduction in the surface band width [93,94]. Theoretically, the band narrowing should be proportional to $\sqrt{N_c}$, in good quantitative agreement with experiment. This assumes a relatively broad cluster size distribution and a negligible interaction between the support and the cluster.

Band narrowing and the observed loss of spectral intensity at the Fermi edge has been interpreted as evidence for an initial state, metal-nonmetal transition. Caution, however, has to be exercised since these spectral changes can arise from final state effects. Recently, based on combined high resolution, low temperature STM/UPS studies, the broadening an positive shift of the Fermi edge of metallic Ag clusters with decreasing size were assigned to delayed neutralization of the photohole [97]. On poorly conducting substrates, the photoelectron experiences a Coulombic attraction on the time scale of the neutralization process that can give rise to the observed shifts. For 2–3 nm clusters on graphite, for example, this neutralization time has been estimated to be on the order of 10^{-15} sec, a time interval on the order of that estimated for tunnel neutralization of an oxide-supported cluster [98]. Unfortunately, the key experiment of varying the velocity of the emitted electron has yet to confirm this. However the shifts observed indicate the importance of the

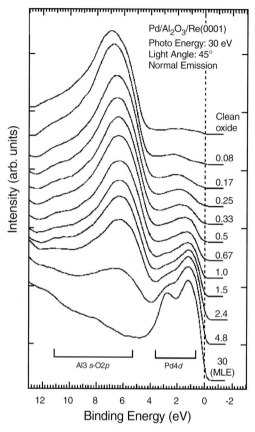

Fig. 7.11. Normal-emission spectra of the valence bands taken at 30 eV photon energy showing with overview of the growth of the Pd clusters on alumina/Re(0001) for coverages ranging from 0.08 to 30 MLE. The binding energy is referred to the Fermi edge of the Re substrate, which is identical to that of thick metallic Pd film at coverage of 30 MLE (from [96])

cluster-support interaction even for femtosecond processes and are particularly relevant to oxide-supported, metal clusters.

The observation of positive core levels shifts for small clusters for those metals which have negative surface-atom core level shifts in the bulk suggests that the change in the surface-to-bulk atom ratio is not a major contributing factor to the observed shifts. The Fermi level shifts have been attributed to Coulombic relaxation, consistent with the positive core level shifts. A simple Coulombic model estimates $\sim 1\,\mathrm{eV}$ for the increase in binding energy for a 1 nm cluster, screened on the femtosecond time scale. Altering the support could decrease this value by a factor of two. Classical electrostatics for the screening dependence of a core hold formed in a metal droplet predicts a $1/R$ behavior for the screening in free and supported clusters.

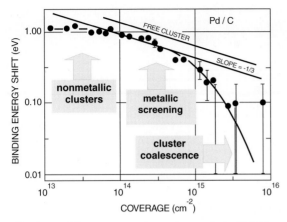

Fig. 7.12. The coverage dependence of the Pd $3d_{5/2}$ electron binding energy (BE) of Pd clusters on a C substrate. The shift with respect to the free clusters is due to substrate polarization and to the nonspherical shape of the supported clusters (from [92])

This $1/R$ behavior has been measured for a number of cluster/support pairs (see Fig. 7.12, for example) including clusters on oxide think films. Final state effects are concluded to be the major factor in the observed energy shifts.

Recently, the generality of this approach and the quantitative predictions have been revised [99]. It was argued that: (i) initial state effects, primarily the lower atom coordination number for small clusters, decrease the cluster electron density (increase the binding energy); and (ii) final state effects, treated as a simple Coulombic interaction of a charged metal droplet with the outgoing photoelectron, are not appropriate in the case of small clusters ($N < 100$) since the conduction band is not developed sufficiently to provide adequate screening. These opposing arguments illustrate the current controversy concerning the basic physics of this important problem (see [100,101]).

It is clear from the above discussion that XPS data alone are insufficient to define the impact of finite size effects on the properties of supported metal clusters. A combination of XPS and Auger spectroscopy, on the other hand, can be especially helpful in separating the contributions to core level shifts [102]. Wagner introduced the so-called Auger parameter, $\alpha = E_b(C) + E_k(CC'C'')$, where $E_b(C)$ is the binding energy for a core level C and $E_k(CC'C'')$ is kinetic energy of electrons belonging to the excitation for the C hole and the Auger relaxation involving the C' and C'' core levels. This parameter is an experimental value that does not depend on the energy reference level or charging of the support. The applicability of core-valence-valence (CVV) transitions is limited and has to be considered separately for every system. It has been shown for clusters (with the assumption of

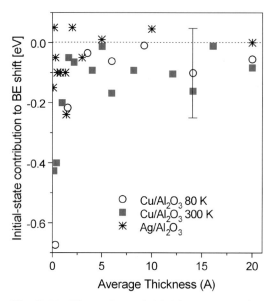

Fig. 7.13. The estimated initial-state contribution to the Cu 2p (312 eV) binding energy shifts (From [104]). For comparison Ag 3d data are plotted as starts (from [105]) (for details see text)

similar core level shifts and similar relaxation for all core levels involved) that the change in the Auger parameter relative to the bulk is approximately twice that of the change in the total relaxation energy, ΔE_r, (the Coulombic term discussed above is included) or $\Delta \alpha = 2\Delta E_r$ [103]. Thus, if the Auger parameter is evaluated, then the initial state contribution, $\Delta \varepsilon$, can be evaluated as $\Delta \varepsilon = \Delta E_b(C) + \Delta \alpha / 2$. For example, in Fig. 7.13, a comparison of the initial state contributions as a function of cluster size for Cu [104] and Ag [105] clusters deposited on alumina thin films are presented. For the smallest Cu clusters, the initial state dominates the core level shift while final state effects dominated ΔE_b for larger clusters. For Ag clusters, final state effects dominate the core level shifts for all clusters sizes. In addition, with similar assumptions (see [106]) and in the absence of initial state effects, $\Delta E_k / \Delta E_b = -3.0$. Thus, a plot of ΔE_b versus ΔE_k should yield a straight line for all clusters sizes as found for the Ag clusters in Fig. 7.14. In agreement with the data shown in Fig. 7.13, the deviation of the slope for Cu at the small cluster limit implies an increasing contribution of initial state effects with decreasing cluster size.

One of the most important size dependent initial state effects is the metal-to-nonmetal transition. This quantum size effect becomes apparent when valence elections are confined within a volume occupied by fewer atom than $N \sim E_f / kT$. Depending on the density of electrons near the Fermi level, N can vary from ~ 200 to 30 for alkali and transition metals, respectively. Kubo's

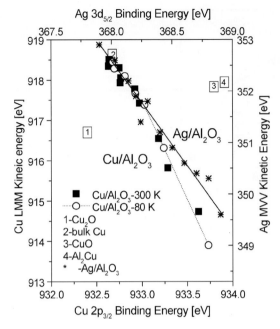

Fig. 7.14. The correlation between CuL$_3$M$_{45}$M$_{45}$ kinetic energies and Cu 2p$_{3/2}$ binding energies for Cu clusters deposited at 300 and 80 K. The values for the compounds are plotted to show the absence of a chemical interaction between the clusters and the support. The *solid line* corresponds to $\Delta KE/\Delta BE = -3.05$, and the *dashed line* indicates the deviation of the relationship at low Cu coverages (adapted from [104]). For comparison Ag 3d$_{5/2}$ and Ag MVV data [105] are plotted in the same graph

approximation [107] is generally supported by valence band photoemission studies on free clusters, and accurately predicts the onset of the metal-insulator transition. One of the first observations of this transition within supported clusters was demonstrated by XPS studies on mass selected small Au clusters deposited on amorphous carbon [108]. In Fig. 7.15 the binding energy for the Au 4f$_{7/2}$ core level is presented as a function of the average coordination number of several cluster sizes with the corresponding bulk values. As decribed above, the general trend is an increase in the binding energy with a decrease in the cluster size (coordinating number). The dashed line approximates the energy shift determined by a $1/R$ Coulombic model, implicating a final state effect. It is evident that the slope of the curve deviates significantly form the expected $1/R$. Figure 7.15 has been assumed to be an indication of the onset of differential screening. The observed transition takes place at a Au cluster size of $N \sim 150$, in agreement with Kubo's predication.

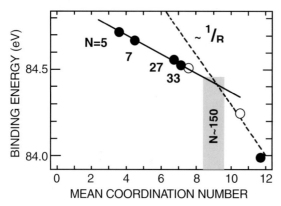

Fig. 7.15. Size selected gold clusters deposited on amorphous carbon. The Au $4f_{7/2}$ binding energy shift as a function of coordination number. The *dashed line* corresponds to the Coulombic final state screening. Point B was measured on the bulk gold and N represents the cluster size. Deviation takes place for clusters with N less than 150 presumably because of the diminution of metallic screening (from [108])

The $1/R$ dependence for the smallest clusters does not guarantee the absence of nonmetallic properties [109]. This dependence was observed for small Rh clusters covered with CO subsequent to their losing their metallic properties and forming Rh carbonyls [110], similar to the behavior of free rare gas clusters [111].

As decribed above, the 2D growth mode of metal clusters often occurs at the early stages of cluster formation on oxides. Being metastable, these 2D clusters exist only in a limited size range of $N \sim 10-100$, yet are catalytically interesting due to their unusual electronic structures and morphologies. Since the metal coverage, Θ, is routinely measured, the slope of a log-log plot of ΔBE versus coverage (Fig. 7.16) can be useful in determining the onset of the 2D-3D transition for small supported clusters. Indeed, if final state effects dominate the binding energy shifts, this slope is proportional to the reciprocal radius of the cluster $\Delta BE \sim 1/R$ (see discussion above). On the other hand, the coverage is related to the dimensionality, D, of the cluster by $\Theta \sim RD$. In Fig. 7.16, similar to that shown in Fig. 7.12, log ΔBE for Pt clusters deposited on thick silica film is plotted as a function of log Θ [112]. The initial stages of cluster growth do not follow the expected $\Theta^{-1/3}$ behavior but rather as $\Theta^{-1/2}$. This, in conjunction with the corresponding XPS intensities, led the conclusion that the 2D-3D transition for Pt clusters occurs at < 1 ML coverage. This is consistent with several other studies on similar systems [89,90].

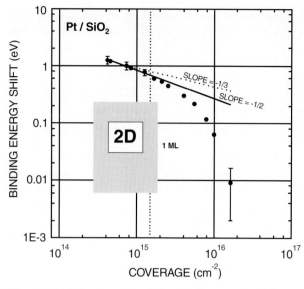

Fig. 7.16. A log-log plot of the binding energy shift versus Pt coverage shows the order of the metal growth, given as the slope. 2D growth is implied by the observed slope of $-1/2$ (adapted from [112])

7.4 Examples of Size Effects in Cluster Reactivity

The following examples were chosen to best demonstrate the typical current structure selectivity research being undertaken on nano-clusters, highlighting new applications and developments of experimental techniques such a scanning tunneling microscopy and spectroscopy, synchrotron radiation based X-ray photoelectron-and absorption spectroscopies, and size selection deposition techniques.

7.4.1 Onset of the Reactivity of Au/TiO$_2$ with Metal-Nonmetal Transitions and the Dimensionality of Supported Clusters

Gold is well known to be one of the most chemically inert metals. However, it has been discovered recently that highly-dispersed gold clusters supported on metal oxides can exhibit extraordinary catalytic activity with respect to many industrially important reactions [113]. The activity of these catalysts and its attenuation with time depend on numerous parameters such as cluster size, the support material, and pretreatment recipes. Due to the complex nature of this behavior, controversy abounds concerning those factors that govern the marked structure selectivity of gold catalysts. These issues have motivated extensive research on gold-based model catalysts. The examples that follow illustrate those undertaken using the surface science approach to this challenging problem, an approach that utilizes a combination of high pressure reactivity measurements coupled with careful characterization of model gold nanoclusters supported on well-ordered oxide surfaces. In particular, electron (XPS), ion scattering (ISS) and scanning tunneling microscopies/spectroscopies have been employed to define those properties of gold nanoclusters supported on TiO_2 that are relevant to their enhanced activity with respect to a particular probe reaction, namely, low temperature CO oxidization.

Planar model catalysts were prepared by controlled vapor deposition of Au onto reduced, bulk $TiO_2(110)$ (n-type) single crystals or on an ultrathin TiO_2 film epitaxially grown on Mo(100). No significant differences were found between these two supports either in deposit morphology or in chemical reactivity. In both cases the conductivity of the support was sufficient for the application of electron/ion spectroscopies and for STM imaging. The Au growth mode on TiO_2 was investigated with ISS and AES, and shown to exhibit a 3D Volmer-Weber growth mode [114]. The surface morphology upon Au deposition is shown schematically in Fig. 7.1. Due to a week Au–TiO_2 interaction, gold forms 3D clusters a few nm in size that nucleated along extended defects or on point defects. However, a detailed analysis of the initial stage of Au cluster growth indicates the existence of a critical coverage, below which 2D clusters are stable (see discussion above and Fig. 7.17) [89,115]. The STM images at a coverage of ~ 0.1 ML (see insertion in Fig. 7.17) confirm the presence of irregular quasi-2D Au islands with an average diameter of about 0.6–2.0 nm and $\sim 1-2$ atomic layers in height, neglecting tip convolution. Assuming a density of bulk Au for these aggregates, the average number of atoms in the cluster is estimated to be $\sim 10-100$. This is precisely the size range within which quantum size effects are expected to become important. As can be seen from the insert on the right of Fig. 7.17, further deposition leads to 3D clusters having 3.0–5.0 nm in diameter and with 1.0–2.5 nm in height, containing on the order of 10^3 atoms at 0.5–1 ML coverage.

No significant Au–TiO_2 interaction for any cluster size probe was determined using XPS. The limited core level shifts, ~ 0.15 eV, detected for Ti 2p and O 1s at submonolayer Au coverages were interpreted as titania downward band bending due to electron transfer from the adsorbate. The negative core level shifts for the smallest Au clusters were attributed to the interplay of both initial and final state cluster size effects.

In addition to the information on cluster morphology, scanning tunneling spectroscopy provides a tool for probing the electronic structure at the nanoscale. The applicability of this spectroscopy for characterization of electronic structure of small supported metal and semiconductor clusters has been demonstrated in a number of studies [116–120]. A closure of the HOMO-LUMO gap with cluster size [121] and the observation of a Coulomb blockage at room temperature [122] are just a few examples of the potential of this

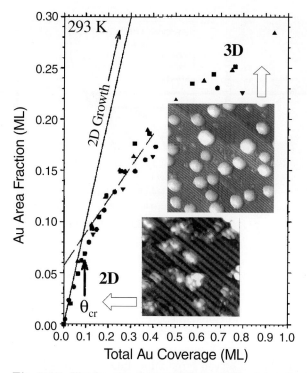

Fig. 7.17. The fractional area of Au on TiO_2 determined from LEIS, as a function of the total Au coverage at 293 K for different Au fluxes from 0.011 to 0.80 ML min^{-1} indicating that the critical coverage is not dependent on the Au flux. The linearity of growth implies 2D islands while the dashed one indicats the formation of the second layer. (From [115].) The inserts are CCT STM images of Au clusters on TiO_2: (*bottom*), a 10×10 nm image of quasi-2D clusters observed after depostion of 0.1 ML Au; and (*top*), a 30×30 nm image of 3D clustes at ~ 1 Au ML coverage

method. However, there are uncertainties in the measurement related to the electronic structure of the tip as well as tip proximity effects. Such effects have to be considered in deconvoluting the cluster electronic structure form the STS data. In addition, since the clusters are supported on poorly conducting oxide films, two tunneling barriers are important: one related to the tip-vacuum-cluster and one related to the cluster-oxide-substrate. However, where tunneling parameters remain constant, clusters are sufficiently large, and when the statistics are good enough the qualitative information related to the development of cluster electronic properties with size acquired with STS is valid. The STS spectra were recorded during topographic imaging by stopping the scan at selected points, sweeping the bias voltage through the region of interest, and measuring the corresponding tunneling current (I) as a function of the bias voltage (V). Fig. 7.18 shows a CCT STM image (*top*) and four STS curves (*bottom*) acquired for clusters of varying

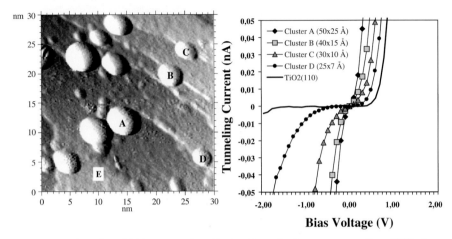

Fig. 7.18. A CCT STM image (2.0 V, 2 nA) and the corresponding STS data acquired for AU clusters of various sizes on the $TiO_2(110)$-(1×1) surface. For reference, the STS curve of $TiO_2(110)$ substrate is also shown (from [125])

sizes on a TiO_2 110 surface. An additional STS (curve E) form the TiO_2 110 substrate is shown for reference. The shape of the I(V) curve and the measured Fermi level position of the TiO_2 substrate are in a good agreement with STS calculations [123]. The observed large tunneling gap is consistent with the semiconducting property of TiO_2 and indicates that the choice of tunneling parameters is adequate to probe the local electronic structure. The I(V) curves measured for various clusters reflect systematically the varying cluster properties, related to cluster size-dependent variations of the local density of states near the Fermi level.

STS curves were obtained for both quasi-2D and 3D clusters. It is noteworthy that the largest change in the I(V) curves and band gap occurs between clusters C and D, where a 2D-3D transition occurs. These results indicate that the nonmetal-metal transition correlates with a dimensionality transition and occurs over a cluster diameter range of 2.0–4.0 nm and a height of ~ 2 atomic layers. The bulk-like electronic structure is well developed at a cluster diameter > 4.0 nm.

In Fig. 7.19b the I(V) characteristics of various Au clusters are shown with respect to their apparent band gaps as a function of cluster size. These data were acquired from $TiO_2(110)$ surfaces with Au coverages ranging from 0.10 to 4.0 ML. Despite the scatter, there is an obvious trend toward a nonzero apparent band gap with a decrease in cluster size. The Au clusters larger than 4.0 nm have fully metallic electronic structures, as seen by the absence of a band gap. The onset of a band gap occurs at a cluster size of ca. 3.5 nm. The apparent band gap continues to increase steadily to > 1.0 V as the cluster size is decreased to 1.5–2.0 nm. On the other hand, the small quasi-2D clusters (< 2.0 nm) have nonmetallic, electronic properties. In Fig. 7.19a the activity

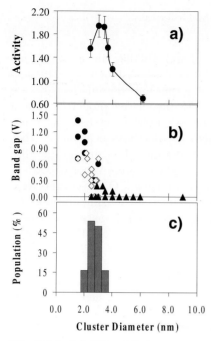

Fig. 7.19. (a) CO oxidation turnover frequencies (TOF = product molecules × (surface atom)$^{-1}$ × (second)$^{-1}$) at 300 K as a function of the average size of the Au clusters supported on a high-surface area TiO_2 support. (From [125].) The *solid line* is drawn simply to guide the eye; (b) cluster apparent band gap measured by STS as a function of the Au cluster size supported on $TiO_2(110)$-(1×1). The band gaps were obtained while acquiring the corresponding topographic scan on various Au coverage ranging from 0.2 ML to 4.0 ML. (●) 2D clusters; (□) 3D clusters with 2 ML in height; (▲) 3D clusters with 3 ML or above in height; and (c) the relative population of the Au clusters exhibiting apparent bad gaps of 0.2–0.6 V as measured by STS

of this catalyst for CO oxidization ($CO:O_2$ = 1:5, P = 40 Torr) at 350 K is presented as a function of Au cluster size. The activity is measured as the number of CO_2 molecules formed per Au atom site. Comparing Figs. 7.19a and 7.19b, a clear correlation is apparent between the band gap measured for the different Au clusters and their corresponding activity. The maximum in the reactivity with respect to cluster size is coincident with the nonmetal-metal transition. The enhanced properties of the gold catalyst decays with time because reaction-induced size deviations of the clusters from the optimum distribution dramatically lowers the overall activity [124]. On the basis of these studies of Au growth, as well as STM/STS and reactivity measurements, the structure sensitivity of CO oxidation on Au/TiO_2 is believed to arise form quantum size effects within the supported Au clusters [125].

7.4.2 CO Dissociation on Structural Defects of Rh/Al$_2$O$_3$/NiAl (110)

It is well known that closed-packed Rh(111) and Rh(100) single crystal surfaces do not dissociate carbon monoxide [126,127], while stepped Rh(210) surface adsorb CO dissociatively [128]. Assuming that low coordination sites are responsible for this enhanced activity, the CO dissociative properties should be even more pronounced for small nanoclusters. Indeed, extensive XPS and XAS studies have shown a marked structural selectivity dependency that maximizes at cluster sizes consisting of ~ 1000 atoms [129]. In these studies 3D Rh clusters from 100 up to 10^5 atoms per cluster were vapor deposited. Subsequent exposure at 90 K to a saturation coverage of CO, the clusters were irradiated with the soft X-ray photon energy tuned for maximum adsorbate sensitivity for XPS and XAS, whereupon data acquired as a function of temperature. Note that in the isothermal spectra the XPS BE's measured for the carbon 1s of CO exhibit a shift toward higher values with decreasing cluster size, attributed by the authors to a final state effect. In Fig. 7.20 the dependence of the intensities of the CO peak ($\sim 286-287$ eV) and the carbon C 1s peak (284 eV) upon heating to 600 K are shown. After heating to 400 K an increase of the dissociatively produced atomic carbon and a decrease of the molecular CO intensity were observed following CO adsorption. The fraction of dissociated CO is plotted in Fig. 7.21 and demonstrates a marked cluster size dependence that maximizes for Rh clusters containing $\sim 10^3$ atoms. In the right panel of Fig. 7.21, the reactivity of stepped Rh(210) and the more densely packed Rh(111) surfaces are presented for comparison. These reactivity with coordination number for CO dissociation.

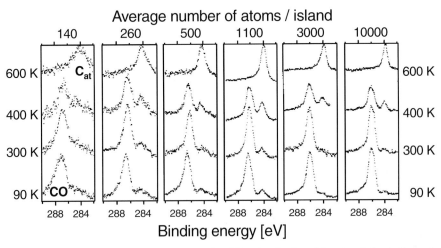

Fig. 7.20. C 1s photoelectron spectra for CO adsorbed on islands of the indicated average number of atoms. The spectra were acquired at 90 K and after annealing the samples to 300, 400, and 600 K (from [129])

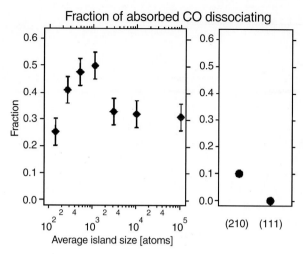

Fig. 7.21. The CO dissociation activity, as determined by comparing the intensities of the C 1s CO features and the final C 1s atomic features (corrected for the 90 K C 1s intensity) for the various conditions. The reaction efficiency for Rh(210) and Rh(111) single crystal surfaces are presented in the graph to the right (from [120])

The temperature-related dissociation of CO suggests a shift of the carbon 1s BE towards lower binding energy with increasing temperature. A careful peak shape analysis of the X-ray data reveals that the complex structure of the CO carbon is feature can be attributed to contributions from two forms of CO adsorbed at two different sites (A and B) (Fig. 7.22). The observed shifts of the CO BE binding energies arise because of the conversion of the A-type species into the B-type species upon heating. That the site corresponding to the B-type CO has a higher coordination number is implied given the decrease in the adsorbate binding BE. This general trend is also size dependent in that its maximum corresponds to the maximum observed in cluster reactivity. It is noteworthy in this respect that the reactivity behavior corresponds to the fraction of B at 300 K (not shown here). As an example, in Fig. 7.23 the temperature evolution for all X-ray data components for a Rh_{500} cluster are shown. Apart from the general trends of CO dissociation and partial desorption above 300 K and the conversion of CO in state A into B, there is a striking asymmetry in temperature dependence between the population of site B CO and atomic carbon. The authors thus justifiable conclude that state B is the precursor to CO dissociation, while state A corresponds to the adsorption site of molecular CO. Accordingly, repeated adsorption of CO leads to population of the A sites exclusively, since the B sites remain blocked with atomic carbon. The exact nature of the A and B sites still remains to be clarified by future investigations. However, the general trend of the cluster size dependence of CO dissociation and the reactivity of stepped surfaces strongly implicates steps as the active sites for CO dissociation. Indeed, with

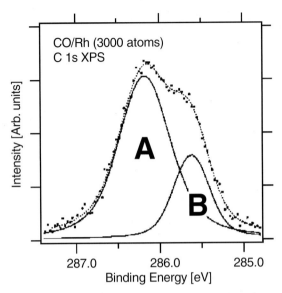

Fig. 7.22. The complex structure of the C 1s feature for a ∼ 3000-atom Rh island after CO adsorption at 90 K, taken with a total resolution of 0.2 eV. The areas and shapes of the components of sites A (desorption sites) and B (dissociation sites) were determined by curve fitting (from [129])

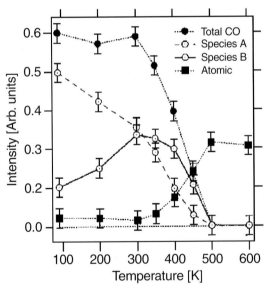

Fig. 7.23. The intensities for the atomic carbon C 1s peak, the CO C 1s peak, and its two components for various heat treatments. The islands consist of ∼ 500 atoms (from [129])

increasing cluster size the equilibrium morphology of the clusters yields facets with less defects and less reactivity.

7.4.3 CO Oxidation Over a Pt/MgO Monodispersed Catalyst

Considerable recent work has been directed toward the production and soft-landing of mass selected clusters onto model supports [44–47]. Because of these efforts, the selectivity of model Pt catalysts has been shown to change with an increase in the cluster size on an atom-by-atom basis [130]. In these experiments positive ions of Pt clusters were formed in a He quenched laser ablation source. Following mass selection the clusters were soft-landed onto a MgO thin film. The total cluster deposition energy was maintained well below the cluster fragmentation threshold. Approximately 10^{14} cluster/cm^2 ($\sim$ 10% of a ML) were deposited at 90 K onto the substrate intact and well-isolated. The nucleation centers for the clusters was presumed to be F-centers and extended defects. Every cluster size was exposed at 90 K to $^{18}O_2$ with an average saturation dose of ca. 20 oxygen molecules per Pt atom. The catalysts were subsequently exposed to the same amount of $^{12}C^{16}O$. Temperature programmed reaction (TPR) was then carried out and the catalytically formed $^{12}C^{16}O^{18}O$ measured. Since the specific size and number of deposited clusters were well known, the number of isotopically labeled CO$_2$ molecules catalytically produced per Pt cluster atom was measured precisely via a calibrated mass spectrometer. (It should be noted that that MgO(100) does not catalyze CO oxidation.)

The evolution of the TPR spectra as a function of cluster size is presented in Fig. 7.24. Several important observations of these experiments are noteworthy: (i) Clusters as small as Pt$_8$ are reactive, and the total reactivity, increasing with increasing cluster size, shows a threshold near $N = 12$ (Fig. 7.25); (ii) with increasing cluster size, the number of reaction sites (labeled as α, β_1, β_2) increases to three for the largest clusters probed. The conclusion that various CO sites are responsible for reaction is confirmed by FTIR spectra (see inserts in Fig. 7.24) suggesting a single CO adsorption site for Pt$_{20}$ clusters and two non-equivalent sites for Pt$_8$ clusters. These varying reaction channels versus size correlates well with established single crystal reaction pathways, i.e.,

$\alpha \longrightarrow O_2^- + 2CO_{terrace} \Rightarrow 2CO_2$ at 160 K on Pt(111) involving hot oxygen atoms and

$\beta_1 \longrightarrow O_{terrace} + CO_{terrace} \Rightarrow 2CO_2$ at 290 K

$\beta_1 \longrightarrow O_{step} + CO \Rightarrow 2CO_2$ at 350 K on a stepped Pt(335) surface.

The presence of the β channels for all clusters sizes is indicative of oxygen dissociation on all the clusters. Notice that the α channel becomes operative only for clusters with $N > 15$. This reaction pathway implies molecular adsorption

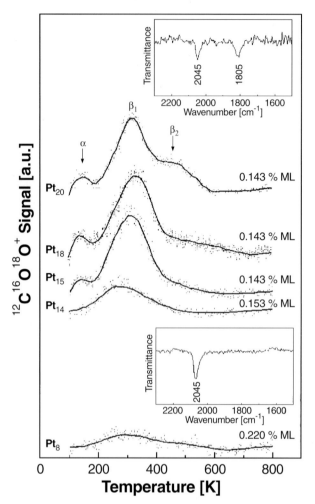

Fig. 7.24. The catalytic yield of CO_2 for various platinum clusters obtained from temperature programmed reaction experiments. The cluster coverage is expressed as percent monolayer, where one monolayer corresponds to 1×10^{15} clusters/cm^2. Various CO_2 formations sites (α, β_1, β_2) are labeled according to single crystal studies. The inserts show the vibrational frequencies of CO adsorbed on Pt_{20} (above) and Pt_8 (below) clusters. (from [130])

when the density of the active low coordinated centers becomes sufficiently small; and (iii) the cluster reactivity normalized per atom indicates a marked atom-by-atom size dependence. This reactivity maximizes at $N = 15$ where 0.4 CO_2 molecules are produced per Pt atom. It should be noted that the reactivity of the Pt(112) atom-fraction is less than 0.2, consistent with the activity trend in Fig. 7.25b.

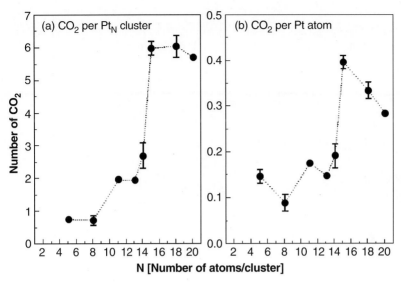

Fig. 7.25. (a) total number of CO_2 molecules produced as a function of clusters size; and (b) the total number of CO_2 molecules produced per atom as a function of cluster size

To describe that general trend in the observed structure-selectivity, the authors proposed that the crucial step for the CO oxidation on a Pt cluster is O_2 dissociation. The oxygen orbitals involved in this process are shown in Fig. 7.26b. According to density functional (DFT) calculations, the position of the Pt d-band with respect to the oxygen π^* state is the most decisive parameter for molecule-surface bonding. Back-donation from the cluster into the antibonding π^* state (or donation from the metal σ/π orbitals into the cluster) is responsible for molecular fragmentation. Since the energy of the HOMO associated with the center of the Pt d-band varies with increasing cluster size (see Fig. 7.26a) there exists an optimum cluster size where there is a resonance between the HOMO of Pt cluster and the π^* of O_2. As can be seen from Fig. 7.26 the clusters with $N \sim 15$ are very close to the resonance conditions in good accord with the observed reactivity results.

An alternative mechanism is related to a possible morphological change of the Pt clusters with increasing size. Free Pt clusters remain planar up to $N = 6$, transitioning to 3D structures for $N > 13$. The authors assumed that the structure of the free clusters was preserved for deposited clusters. With this assumption, the variation of cluster reactivity coincides with a 2D-3D transition in the range of $N = 6-13$. Note also the reactivity minimizes at Pt_{13}, which is presumably related to atomic-shell filling of this icosahedral (or cuboctahedral) cluster. The reduction of the reactivity per atom for $N > 15$

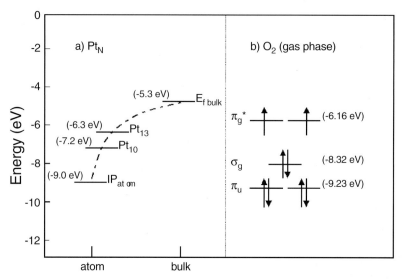

Fig. 7.26. An energy diagram of the relevant electronic states for oxygen dissociation in gas phase clusters and free oxygen. The cluster HOMO lies between the atomic limit (-9.00 eV) and the bulk limit (-5.32 eV). Dissociation is favored if the energy of the cluster's HOMO is close to the antibonding π^* state of oxygen. The dashed curve corresponds to that of the classical conductive droplet. (from [130])

may be related to a decrease in the relative number of active centers with an increase in the total number of Pt atoms.

7.5 Concluding Remarks and Future Prospects

Size dependent cluster reactivity is critical to our understanding of finite-size effects in supported metal catalysis and for developing novel, new industrial catalysts based on these concepts. The synthesis of nanocluster catalysts, with predictable reactivity and selectivity, by tuning the cluster size, composition and morphology is a goal of applied catalytic science. There is a general consensus that quantum size effects play an important role in defining the properties of clusters below 1–2 nm in diameter. Other factors, such as reduced coordination number and specific sites/facets, dominate structure sensitivity for the larger clusters. The influence of the support on cluster reactivity remains the most difficult issue to address. To bridge the so-called "material gap", thin oxide films used as model supports have been implemented. Support characterization, the role of surface defects, electronic structure, and cluster reactivity are subjects to intensive current interest. Mass selected cluster deposition using modern cluster beam techniques is now being widely implemented. This has allowed a comparison of gas phase mass selected clusters and the corresponding supported clusters, thus providing new insights into the role of the support in altering the catalytic properties of metal clusters.

The application of surface science techniques toward the study of clusters has been extremely fruitful during the last two decades. However, much remains to be understood regarding the relationship between cluster electronic properties and cluster size, the interaction of the cluster with the support, and cluster reactivity. In addition to the extensive use of conventional laboratory surface techniques and scanning probe microscopies, synchrotron-based studies will be increasingly more important in the coming years. Recent developments in various synchrotron-based spectroscopies and microscopies will permit the exploration of the local electronic, geometrical and chemical properties of clusters at the nanoscale. Finally the relatively new field of real time monitoring of heterogeneously catalyzed chemical reactions on nanoscale systems is currently being developed and refined.

Acknowledgements. We acknowledge with pleasure the support of this work by the Department of Energy, Office of Basic Energy Sciences, Division of Chemical Sciences and the Robert A. Welch Foundation. We also thank J. Stultz for his contributions in the preparation of the manuscript.

References

1. R.L. Whettern, D.M. Cox, D.J. Trevor, A. Kaldor, Surf. Sci. **156** (1985); D.M. Cox, A. Kaldor, P. Fayet, W. Eberhardt, R. Brickman, R. Sherwood, Z. fu, D. Sondericher, ACS Symposium Series 437 (1990) 172
2. S.J. Riley in "Clusters of Atoms and Molecules II: Supported and Compressed Clusters", H. Haberland (ed.), Berlin, Springer-Verlag (1994) 221
3. M.B. Knickelbein, Annu. Rev. Phys. Chem. **50** (1999) 79
4. M. Ichikawa in "Chemisorption and Reactivity on Supported Clusters and Thin Films" ed. by R.M. Lambert and G. Pacchioni, NATO ASI Series 331 (1997) 153.
5. C.R. Henry, C. Chapon, S. Giorgio, C. Goyhenex in "Chemisorption and Reactivity on Supported Clusters and Thin Films" ed. by R.M. Lambert and G. Pacchioni, NATO ASI Series **331**, 117 (1997).
6. M. Che, C.O. Bennett, Adv. Cat. **36** (1989) 55
7. D.W. Goodman, surf. Rev. Lett. **2**,(1995) 9
8. P.L.J. Gunter, J.W.H. Niemantsverdriet, F.H. Ribeiro, G.A. Somorjal, Catal. Rev. Sci. Eng. **39**, 77 (1997).
9. H.-J. Freund, Angew. Chem. Int. Ed. Engl. **36**, 452 (1997).
10. C.T. Campbell, Surf. Sci Rep. **27**, 1 (1997).
11. C.R. Henry, Surf. Sci. Rep. **31**, 235 (1998).
12. M. Baumer and H.-J. Freund, Prog. Surf. Sci. **61**, 127 (1999).
13. U. Heiz and W.-D. Schneider, J. Phys. D Appl. Phys. **33**, R85 (2000).
14. D.R. Rainer and D.W. Goodman, J. Mol. Cat. A **131**, 259 (1998).
15. T.P.St. Clair, D.W. Goodman, Top. Cat. **13**, 5 (2000).
16. R. Persaud, T.E. Madey in : "Growth and Properties of Ultrathink Epitaxial Layers", D.A. King, D.P. Woodruf (Eds.), Elsevier, Amsterdam (1997) 407

17. U. Diebold, J.-M. Pan, T.E. Madey, Surf. Sci. 331-333, 845 (1995).
18. R. Franchy, Surf. Sci. Rep. **38** (6–8) (2000) 199; C. Xu and D.W. Goodman in "Handbook of Heterogeneous Catalysis", G. Ertl, H. Knotzinger and J. Weikamp (Eds.), Wiley-VCH, Weinheim **2**, 826 (1997).
19. H.-J. Freund, Phys. Status Solidi (b) **192**, 407 (1995).
20. H.-J. Freund, Faraday Disc. **114**, 1 (1999).
21. J.G. Chen, J.E. Crowell, J.T. Yates Jr., Surf. Sci. **185**, 373 (1987).
22. S.T. Pantelides (Ed.), "The Physics of SiO_2 and its Interfaces", Pergamon Press, New York (1978)
23. R.M. Jaeger, H. Kuhlenbeck, H.-J Freund, M. Wuttig, W. Hoffmann, R. Franchy, H. Ibach, Surf. Sci. **259**, 235 (1991) 235
24. J. Libuda, F. Winkelmann, M. Baumer, H.-J. Freund, Th. Bertrams, H. Neddermeyer, K. Muller, Surf. Sci. **318**, 61 (1994).
25. D.W. Goodman, J. Vac. Sci. Technol. A **14**, 1526 (1996).
26. H. Poppa, Cat. Rev. Sci. Eng. **35**, 359 (3) (1993).
27. V.E. Henrich and P.A. Cox, "The surface science of metal oxides", Cambridge University Press (1994)
28. J. Libuda, M. Frank, A. Sandell, S. Andersson, P.A. Bruhwiler, M. Baumer, N. Martensson, H.-J. Freund, Surf. Sci. **384**, 106 (1997).
29. R. Kern in "Morphology of Crystals" I. Sunagawa (ed.), Terra, Tokyo, (1987) 77
30. A. Pinto, A.R. Pennisi, G. Faraci, G. D'Agostino, S. Mobilio, F. Boscherini, Phys. Rev. B **51**, 5315 (1995).
31. K.H. Ernst, A. Ludviksson, R. Zhang, J. Yoshihara, C.T. Campbell, Phys. Rev. B **47**, 13782 (1993).
32. L. Zhang, R. Persaud, T.E. Madey, Phys, Rev. B **56**, 10549 (1997).
33. X. Lai, T.P.St. Clair, M. Valden, D.W. Goodman, Prog. Surf. Sci. **59**, 25 (1998).
34. W. Hoheisel, K. Jungmann, M. Vollmer, R. Weidenauer, F. Trager, Phys. Rev. Lett. **60**, 1649 (1988).
35. J. Bosbach, D. Martin, F. Stietz, T. Wenzel, F. Trager, Eur. Phys. J. D (9) (1-4) (1999) 613; T. Wenzel, J. Bosbach, A. Goldmann, F. stietz, F. Trager, Appl. Phys. B-Lasers **69** (5–6), 513 (1999).
36. U. Kreibig and M. Vollmer, "Optical Properties of Metal Clusters", Springer Ser. Mater. Sci. Vol. **25**, Springer-Verlag, Berlin (1995)
37. "Clusters and Colloids: from theory to applications", Gunter Schmid (Ed.), Weinheim, New York, VCH (1994)
38. L. Rapoport, Y. Bilik, Y. Feldman, M. Homyonfer, S.R. Cohen, R. Tenne, Nature **387**, 791 (1997).
39. E.E. Carpenter, A. Kumbhar, J.A. Wiemann, H. Srikanth, J. Wiggins, W.L. Zhou, C. J. O'Connor, Mater. Sci. Eng. A **286**, 81 (2000).
40. R. Wilkins, E. ben-Jacob, R.C. Jaklevic, Phys. Rev. Lett. **63**, 801 (1989).
41. R.P. Andres, T. Bein, M. Dorogi, S. Feng, J.I. Henderson, C.P. Kubiak, W. Mahoney, R.G. Osifchin, R. Reifenberger, Science, **272**, 1323 (1996); R.P. Andres, J.D. Bielefeld, J.I. Henderson, D.B. Janes, V.R. Kolagunta, C.P. Kubiak, W.J. Mahoney, R.G. Osifchin, ibid. **273**, 1690 (1996).
42. G. Schmid, M. Baumle, M. Geerkens, I. Heim, C. Osemann, T. Sawitowski, Chem. Soc. Rev. **28**, 179 (1999).
43. F. Vanolli, U. Heiz, W.-D. Schneider, Surf. Sci **414**, 261 (1998).

44. K. Bromann, C. Felix, H. Brune, W. Harbich, R. Monot, J. Buttet, K. Kern, Science **274**, 956 (5289) (1996).
45. W. Harbich, Philos. Mag. B **79** (9), 1307, (1999).
46. U. Heiz, F. Vanolli, L. Trento, W.-D. Schneider, Rev. Sci. Instr. **68**, 1986, (1997).
47. P.W. Jacobs, S.J. Wind, F.H. Ribeiro, G.A. Somorjai, Surf. Sci. **372** (1–3) L249, (1997).
48. M.X. Yang, D.H. Gracias, P.W. Jacobs, G.A. Somorjai, Langmuir **14**, 1458 (1998).
49. B. Kasemo, S. Johansson, H. Persson, P. Thormahlen, V.P. Zhdanov, Top. Cat. **13** (1–2), 43 (2000).
50. V.P. Zhdanov, B. Kasemo, Surf. Sci. Rep. **39** (2–4), 29 (2000).
51. A. Avoyan, G. Rupprechter, A.S. Eppler, G.A. Somorjai, Top. Cat. **10** (1–2), 107 (2000).
52. Z.X. Yang, R.Q. Wu, D.W. Goodman, Phys. Rev. B **61** (20), 14066 (2000).
53. K. Luo, T.P.St. Clair, X. Lai, D.W. Goodman, J. Phys. Chem. B **104** (14), 3050 (2000).
54. H.-J. Freund, M. Bäumer, H. Kuhlenbeck, Adv. Cat. **45**, 333 (2000).
55. U. Heiz and W.-D. Schneider in "Metal clusters at Surfaces" ed. by K.-H. Meiwes-Broer, Springer (2000).
56. R. Denecke, P. Vaterlein, M. Bassler, N. Wassdahl, S. Butorin, A. Nilsson, J.-E. Rubensson, J. Nordgren, N. Mårtensson, R. Nyholm, J. El. Spec. Rel. Phen. **103**, 971 (1999).
57. See for example J. Stöhl, "NEXAFS Spectroscipy", Springer-Verlag, Berlin (1992).
58. See for example B.K. Teo, "EXAFS-basic principles and data analysis", springer, NY (1986).
59. G. Sankar and J.M. Thomas, Top. Cat. **8** (1,2), 1 (1999).
60. D.C. Koningsberger, B.I. Mojet, G.E. Van Dorssen, D.E. Ramaker, Top. Cat. **10**, 143 (2000).
61. D. van Heijnsbergen, G. von Helden, M.A. Duncan, A.J.A. van Roij, G. Meijer, Phys. Rev. Let. **83**, 4983 (1999).
62. R.J. Imbihl, Mol. Cat. A **158** (1), 101 (2000).
63. E. Taglauer, H. Knozinger, S. Gunther, Rev. Sci. Instr. **158** (1–4), 638 (1999).
64. T. Warwick, H. Ade, S. Cerasari, J. Denlinger, K. Franck, A. Garcia, S. Hayakawa, A. Hitchcock, J. Kikuma, S. Klingler, J. Kortright, G. Morisson, M. Moronne, E. Rightor, E. Rotenberg, S. Seal, H.-J. Shin, W.F. Steele, B.P. Tonner, J. Synchr. Rad. **5**, 1090 (1998).
65. M.P. Kiskinova, Surf. Int. Analys. **30** (1), 464 (2000).
66. E. Bauer, Rep. Prog. Phys. **57**: (9), 895 (1994). Also: b. P. Tonner, D. Dunham, T. Droubay, J. Kikuma, J. Denlinger, J. El. Spec. **78**, 13 (1996).
67. T. Schmidt, S. Heun, J. Slezak, J. Diaz, K.C. Prince, G. Lilienkamp and E. Bauer, Surf. Rev. Lett. **5**(6), 1287 (1998).
68. R. Wichtendahl, R. Fink, H. Kuhlenbeck, D. Preikszas, H. Rose, R. Spehr, P. Hartel, W. Engel, R Schlögl, H.-J. Freund, A.M. Bradshaw, G. Lilienkamp, T. Schmidt, E. Bauer, G. Benner and E. Umback, Surf. Rev. Lett. **5**(6), 1249 (1998). Also: J. Stöhr, S. Anders, IBM J. Res. Develop. **44** (4), 535 (2000).
69. G.A. Somorjai, Appl. Surf. Sci. **121/122**, 1 (1997).
70. G. Ertl and H.-J. Freund, Phys. Today (1999) 32.

71. L.J. Lauhon, W. Ho, Phys. Rev. Lett. **84**, 1527 (2000).
72. D.A. Bonnell, Prog. Surf. Sci. **57** (3), 187 (1998).
73. J. Viernow, D.Y. Petrovykh, A. Kirakosian, J.-L. Lin, F.K. Men, M. Henzler, and F.J. Himpsel, Phys. Rev. B **59**, 10356 (1999).
74. D.L. Carroll, M. Wagner, M. Ruhle, D.A. Bonnell, Phys. Rev. B **55**, 9792 (1997).
75. E. Perrot, A. Humbert, A. Piednoir, C. Chapon, C.R. Henry, Surf. Sci. **445**, 407 (2000).
76. K. Højrup Hansen, T. Worrew, S. Stempel, E. Lægsgaard, M. Bäumer, H.-J. Freund, F. Besenbacher, I. Stensgaard, Phys. Rev. Let. **83**, 4120 (1999).
77. R.A. Bennett, P. Stone, M. Bowker, Fraday Discuss. **114**, 267 (1999).
78. O. Dulub, W. Hebenstreit, U. Diebold, Phys. Rev. Lett. **84** (16), 3646 (2000).
79. J.A. Jensen, K.B. Rider, Y. Chen, M. Salmeron, G.A. Somorjai, J. Vac. Sci. Tech. B **17** (3), 1080 (1999).
80. A. Kolmakov and D.W. Goodman, Surf. Sci. **490**, L597 (2001).
81. P.L. Gai, Top. Cat. **8**, 97(1999).
82. T. Buffeteau, B. Desbat and J.M. Turlet, Appl. Spectr. **45**, 380 (1991).
83. H. Knözinger and G. Mestl. Top. Cat. **8**, 45(1999) 45.
84. M.J. Weaver, Top. Cat. **8**, 65 (1999).
85. G.A. Somorjai, X.C. Su, K.R. McCrea, K.B. Rider, Top. Cat. **8**, 23 (1999).
86. M. Haruta, Stud. Surf. Sci. Catal. **110**, 123 (1997).
87. S.A. Mepijko, M. Klimenkov, H. Kuhlenbeck, D. Zemlyanov, D. Herein, R. Schlögl, H.-J. Freund, Surf. Sci. **413**, 192 (1998).
88. P. Muller, R. Kern, Appl. Surf. Sci. **164**, 68 (2000); Also Surf. Sci. **162**, 133 (2000).
89. K.H. Ernst, A. Ludviksson, R. Zhang, J. Yoshihara, C.T. Campbell, Phys. Rev. B **47**, 13782 (1993).
90. L. Zhang, R. Persaud, T.E. Madey, Phys. Rev. B **56**, 10549 (1997).
91. R.L. Whetten, D.M. Cox, D.J. Trevor, A. Kaldor, Phys. Rev. Lett. **54**, 1494 (1985).
92. G.K. Wertheim, S.B. DiCenzo, D.N.E. Buchanan, Phys. Rev. B **33**, 5384 (1986).
93. G.K. Wertheim, S.B. DiCenzo, S.E. Youngquist, Phys. Rev. Lett. **51**, 2310 (1983).
94. G.K. Wertheim, Z. Phys. D **12**, 319 (1989).
95. S.B. DiCenzo, G.K. Wertheim, Comments Solid State Phys. **11**, 203 (1985).
96. Y.Q. Cai, A.M. Bradshaw, Q. Guo, D.W. Goodman, Surf. Sci. **399** (2-3), L357 (1998).
97. H. Hövel, B. Grimm, M. Pollmann, and B. Reihl, Phys. Rev. Lett. **81**, 4608 (1998).
98. A. Sandell, J. Libuda, P.A. Brühwiler, S. Andersson, M. Bäumer, A.J. Maxwell, N. Mårtensson, H.-J. Freund, Phys. Rev. B **55**, 7233 (1997).
99. P.S. Bagus, F. Illas, G. Pacchioni, F. Parmigiani, J. El. Spec. Rel. Phenom. **100**, 215 (1999).
100. F. Parmigiani, E. Kay, P.S. Bagus, C.J. Nelin, J. El. Spec. Rel. Phenom. **36**, 257 (1985).
101. P.S. Bagus, C.J. Nelin, E. Kay, F. Parmigiani, J. El. Spec. Rel. Phenom. **43**, c13 (1987).
102. B.D. Wagner, Faraday Discuss. Chem. Soc. **60**, 110 (1975).

103. G.K. Wertheim, Phys. Rev. B **36**, 9559 (1987).
104. Y. Wu, E. Garfunkel and T.E. Madey, J. Vac. Sci. Tech. A **14** (3), 1662 (1996).
105. K. Luo and D.W. Goodman (in preparation)
106. J. Jirka, Surf. Sci. **232**, 307 (1990).
107. R. Kibo, A. Kawabata, S. Kabayashi, Ann. Rev. Mater. Sci. **14**, 49 (1984).
108. S.B. DiCenzo, S.D. Berry, E.H. Hartford Jr., Phys. Rev. B **38**, 8465 (1988).
109. O.D. Häberlen, S.-C. Chung, M. Stener, N. Rösch, J. Chem. Phys. **106**, 5189 (1997).
110. C.F. Frederick, G. Apai, T.N. Rhodin, J. Am. Chem. Soc. **109**, 4797 (1987).
111. F. Federmann, O. Bjorneholm, A. Beutler, T. Möler, Phys. Rev. Lett. **73** (11), 1549 (1994).
112. J.W. Keistler, J.E. Rowe, J.J. Kolodziej, T.E. Madey, J. Vac. Sci. Tech. B (18), 2174 (2000).
113. M. Haruta, Catal. Today **36**, 153 (1997).
114. X. Lai, T.P.St. Clair, M. Valden, D.W. Goodman, Prog. Surf. Sci. **59** (1-4), 25 (1998).
115. S.C. Parker, A.W. Grant, V.A. Bondzie, C.T. Campbell, Surf. Sci. **441** (1), 10 (1999).
116. R.M. Feenstra, Phys. Rev. Lett., **63**, 1412 (1989).
117. A. Bettac, L. Koeller, V. Rank, K.-H. Meiwes-Broer, Surf. Sci. 402-404 (1998) 475.
118. M. Lonfat, B. Marsen, K. Sattler, Chem. Phys. Lett. **313**, 539 (1999).
119. B. Xu, D.W. Goodman, Chem. Phys. Lett. **263** (1-2), 247 (1996).
120. C. Xu, X. Lai, D.W. Goodman, Faraday Discuss. **105**, 247 (1996).
121. P.N. First, J.A. Stroscio, R.A. Dragoset, D.T. Pierce, R.J. Celotta, Phys. Rev. Lett. **63** (13), 1416 (1989).
122. D. Schönenberger, H. van Houten, H.C. Donkersloot, Europhys. Lett. **20**, 249 (1992).
123. E.A. Bonnell, J. Am. Ceram. Soc. **81** (12), 3049 (1998).
124. M. Valden, S. Pak, X. Lai, D.W. Goodman, Catal. Lett. **56** (1), 7 (1998).
125. M. Valden, X. Lai, D.W. Goodman, Science **281**, 1647 (1998).
126. Y. Kim, H.C. Peebles, J.M. White, Surf. Sci. **114**, 363 (1982).
127. J.T. Yates Jr., E.D. Williams, W.H. Weinberg, Surf. Sci. **91**, 562 (1980).
128. M. Rebholz, R. Prins, N. Kruse, Surf. Sci. Lett. **259**, L797 (1991).
129. S. Andersson, M. Frank, A. Sandell, A. Giertz, B. Brena, P.A. Brühwiler, N. Mårtensson, J. Libuda, M. Baümer, H.-J. Freund, J. Chem. Phys. **108**, 2967 (1998).
130. U. Heiz, A. Sanchez, S. Abbet, W.-D. Schneider, J. Am. Chem. Soc. **121**, 3214 (1999).

8 Delayed Ionization

E.E.B. Campbell and R.D. Levine

Long-delayed ionization is a statistical phenomena governed by the density of states. Like other statistical processes, its rate is governed by the magnitude of Planck's constant. We also discuss newer work that identifies more prompt ionization channels, channels that are equally not direct. The existence of several time scales for ionization is discussed both quantum mechanically and statistically. Experimental results for fullerenes are presented.

8.1 Introduction

Hot systems with sufficient internal energy can emit electrons. There are other endoergic decay channels which can also cool hot systems (bond breaking, radiative emission, ...) and, if these channels dominate, they can overwhelm cooling which proceeds via ionization. Hence delayed ionization is experimentally important when the energy threshold for ionization is low and/or the atomization energy is high. Therefore, bulk metals provided the paradigm for thermionic emission. The computation (by Richardson, in 1920 [1–3]) of the rate of electron emission is the earliest known example of a formulation of a statistical theory of a rate process and is a forerunner of transition state theory. More recently, delayed ionization has been observed in a variety of refractory clusters and nanosystems [4–25] and also for weakly bound electrons in ordinary clusters (e.g. [26]). A new development is the observation of non-prompt ionization following selective excitation, e.g. [27–32]. In many, but not all [28,29], such processes the time scale for ionization is shorter than for thermionic emission, but in all these examples ionization is not prompt; rather, it is delayed to a smaller or larger extent. In this chapter we deal first with the long-delayed ionization that, by analogy to the bulk process, often goes under the name of thermionic emission and then discuss the dynamics of shorter-time but still not prompt processes, with experimental results for fullerenes. For a more complete literature coverage, arranged by systems and mode of excitation and additional material see [33]. For a more general discussion of several time scales in delayed ionization and other unimolecular processes see [34,35]. In particular, a review of the quantum theory of delayed vs. prompt dynamics is available in [36]. Here we limit consideration to a kinetic model that captures the essence of the quantum mechanical description.

Delayed ionization occurs on the μs time scale. By such a long time one expects that the hot system has representatively sampled its available phase space. This is a characteristic of systems in thermal equilibrium and so the process is known as thermionic emission. As is well understood, representative sampling of the available phase space is subject to conservation of energy (and other good constants of the motion). Therefore the system need not really be thermal. The rate of statistical ionization needs to be computed at a given total energy. If this is called for, the rate can then be averaged over the actual distribution of the excitation energy of the molecule. This distribution depends on the preparation process. It can be thermal but it need not be and, at the risk of belaboring the obvious, no amount of waiting can modify the, initially determined distribution of the total energy. Waiting can deplete the amount of unionized (or unfragmented, etc.) higher energy systems, unlike a bulk sample of a metal which is continuously in contact with a heat bath so that its energy distribution is maintained despite cooling by electron emission. The section on transition state theory provides the essentials of computing the rate of ionization at a given total energy. If the excitation process has not prepared a nearly uniform distribution in the available phase space or, if time has not erased the details of the initial preparation, then one needs a more elaborate approach. One of the implications of this will be the existence of several time scales for the decay of the energy rich molecule.

That ionization of metals can be rather prompt has been recognized for over 100 years: Early studies of the photoelectric effect failed to discern any delay between the incident light and the emitted electrons. These experiments also reported that light at too long a wavelength, however intense (by 19th century standards), failed to induce a photocurrent. The modern version of such experiments is what we know as photoelectron spectroscopy and the photoelectrons are indeed prompt. So we are equally familiar with both prompt ionization, where the excitation energy needs to be resonant with a molecular transition and long delayed ionization, which is statistical without energy selectivity. The last part of this chapter will emphasize that there is an intermediate regime. The experimental evidence is clear and abundant. Specifically, if we consider a molecule, even a diatomic will do, photoexcited above the lowest threshold for ionization, and we wait a while so that the prompt electrons move out of the detection region, the electron emission does not stop. Nor does the non-prompt emission occur on a single time scale [29,30].

8.2 Transition State Theory

We consider first an energy rich molecule with a total energy in the range $E, E + \delta E$ where δE is a small increment. When E exceeds the ionization potential I, the molecule can ionize. It will not necessarily ionize immediately because, if the energy is distributed over many vibrational and electronic modes, it takes time before a fluctuation localizes energy in excess of I. We take it that this energy is to be localized in one coordinate and that one can define a critical location along this coordinate such that, once the departing electron passes this point, it will certainly depart. This is the point of no return. The assumption that such a point can be defined is the essential approximation made by the theory. Unlike the usual situation in chemical kinetics where there is a barrier along the reaction coordinate, here the situation is not quite as simple. In the original formulation of the rate of thermionic emission, one placed an imaginary surface parallel to the face of the metal and displaced just a shade into the vacuum. On the vacuum side it is assumed that there is no potential seen by the electron. On the metal side the electron is bound by at least the work function. Crossing this surface is taken to be the act of no return. It is less obvious how to define such a surface for a molecular system. The problem is most noticeable for ionization of neutral species. The long range potential between the electron and the positive core that is left behind is Coulombic. It is an attractive potential without any hump. In reality there is always an external field and the presence of this additional field does lead to a clear saddle point (e.g. [37]). In this case an outgoing electron can be turned back by the coupling to the ionic core. For detachment from a negative ion, where the remaining core is neutral, the long range attractive potential scales as R^{-4} and there is a maximum in the potential when one includes the orbital angular momentum of the electron. For clusters or for C_{60} it has become customary to use the geometrical shape (i.e., a spherical surface of no return), by analogy to the plane separating the metal from the vacuum.

There is a second ingredient that goes into the theory. It is that all states of the energy rich system are equally probable. The validity of this assumption is critically dependent on the method of excitation. For example, if the system is truly thermal, then this condition is inherently implied by the system being in thermal equilibrium. That is why, for thermal bimolecular reactions, transition state theory has only one assumption, namely that there is a point of no return. Our case is more complicated because most methods of excitation do not prepare a thermal distribution. In fact, we do not need a thermal distribution. It is sufficient if a given total energy is uniformly distributed. This is the analogue of the RRKM assumption of unimolecular dissociation. But it is a stronger assumption because we need the electronic degrees of freedom to also be in equilibrium and to be in equilibrium with the vibrational modes. Given a sufficient delay after the excitation, such an equilibrium will prevail but during the long wait there is time for other decay processes to take place with the result that very few systems survive to exhibit statistical behavior. The kinetic model to be discussed below will emphasize this point. For the moment, we assume that the excitation process is sufficiently non selective or/and that enough time has passed that we can assume that all states at a given narrow energy range $E, E + \delta E$ are equally probable.

The potential energy of the system at the point of no return is denoted E_0. States can promptly ionize when energy in excess of E_0 is available at the point of no return. Some of the excess energy, $E - E_0$, is taken by the departing electron and the rest is an excitation of the core. To compute the total rate of ionization we will first determine the rate for each state of the core when the energy is in the range $E, E + \delta E$. The result will turn out to be remarkably simple namely, each such quantum state ionizes with the rate $(1/h)\delta E$ where h is Planck's constant. This will be shown below and it is in this result that Planck's constant enters in an essential way, that sets the time scale. Next, the total rate is the sum of the rates of all states that can promptly ionize. Since all states are, by assumption, equally occupied, the total rate is the rate per state, $(1/h)\delta E$, times the number, $N_\ddagger$, of states of the core. Note how the assumption of uniform distribution of the excess energy was used in an essential way. If this assumption fails, and different states have different weights, then the total rate is a weighted sum where each state is given its correct weight and then we need to know these weights.

To derive the rate of ionization we use a semiclassical argument. Consider a small length range δs along the coordinate of the departing electron, spanning the point of no return. At that point the electron is not strongly coupled to the core so a state of the system can be factorized as a state of the core and a state of the electron. The available energy, $E - E_0$, is the sum of the (quantized) internal energy of the core and the kinetic energy ε of the electron. The momentum p of the electron is determined by ε, $\varepsilon = p^2/2m$. The core is so heavy that m is the mass of the electron. The small but finite range of E means that there is a corresponding range in ε and hence in p, $\delta\varepsilon = p\,dp/m = \nu\,dp$ where ν is the velocity of crossing the point of no return. The number of quantum states for the electron in the narrow interval spanning the point of no return is $ds\,dp/h$. The number density of quantum states is dp/h and the rate of crossing (= the flux density) is $\nu\,dp/h = d\varepsilon/h = dE/h$.

In the derivation in the paragraph above, the state of the core was fixed and thereby the energy and hence the velocity of the outgoing electron were confined to a narrow range. We computed the rate for all possible states of the electron that go with the given state of the core. All these states of the electron were taken to be equally probable. Next we complete the summation over all quantum states of the entire system by summing over all states of the core. These states of the core range from the ground state, when all the available energy, $E - E_0$, is the kinetic energy of the departing electron up to states with excitation energy $E - E_0$ so that the electron barely crawls across the point of no return. Each one of these states of the core contributes the very same flux of outgoing electrons so that the total rate of ionization

(= total flux or, when multiplied by the electronic charge e, the current) is

$$I(E)\delta E = e\frac{N_{\ddagger}(E - E_0)}{h}\delta E . \tag{8.1}$$

Here $N_{\ddagger}(E - E_0)$ is the number of states of the core that are accessible at the point of no return. Two points need to be emphasized: The term in brackets is the argument of the function, 'the number of states of the core with an excitation energy up to x', and that it is the number of states of the remaining core not the number of states of the system that is being counted. The subscript $\ddagger$ is a reminder of this meaning of the number of states.

In principle, this is it. We have reduced the problem of computing the rate of delayed ionization of a molecule at the energy E to the far simpler problem of counting how many states of the core can be accessed. In practice, the required counting is not so simple. Why? We do have a range of options for determining $N_{\ddagger}$ starting with the simple RRK estimate all the way to computer programs that do state counting. The problem is the high energies that are typical of systems that can ionize. At such energies it is just not enough to count the vibrational states of the ground isomer. Larger polyatomic systems have numerous higher energy geometrical isomers, each with its complement of states. For metallic clusters there is the problem of excited electronic states with the limiting case of a bulk metal where all the states that are counted in the Richardson formula are electronic.

The rate constant is defined as the average over the initial states. The current is the sum over all initial states. Since each term in the sum is identical, we can get the rate constant by dividing the total rate of ionization (= current/e) by the number of initial states. When the energy is in the range $E - E_0$ the number of states is $\rho(E)\delta E$ where $\rho(E)$ is the density of states. So the rate constant for ionization at the energy E is given by $k(E) = e^{-1}I(E)\delta E/\rho(E)\delta E$ or, explicitly

$$k(E) = \frac{N_{\ddagger}(E - E_0)}{h\rho(E)} . \tag{8.2}$$

8.3 Detailed Balance

Detailed balance is a condition that the rate constant needs to satisfy. It is an exact condition and, when applied correctly, detailed balance offers a powerful tool for relating different processes. Some people are reluctant to see detailed balance being used because sometimes the claims for what it can do are too strong. The point is that while detailed balance is a necessary condition in a well-founded theory, satisfying detailed balance does not imply that the result is exact. An example is transition state theory. It strictly satisfies detailed balance. This is quite easy to show and we will provide the proof below. Yet (8.2) is not an exact result for the rate constant. Other people are reluctant to see detailed balance being used because its correct application requires a careful definition of the state of the reactants in the processes that are being related. To avoid misunderstanding, our discussion will therefore be pedantic. We apologize but the alternative is to leave room for ambiguity. One thing that detailed balance does *not* require is the determination of the quantum state(s) of the products. There is *no* assumption that the reaction proceeds to a statistical (or any other) product state distribution. There is no implication that the nascent product state distribution needs to be the same as the distribution of the reactants of the reversed reaction. One of the benefits of a pedantic approach is that it makes this point quite clearly.

Detailed balance can be derived from the quantal (or classical) equations of motion. A complementary derivation comes from the condition that the system reaches a dynamical equilibrium. By 'dynamical' we mean that, at equilibrium, it is not the case that all processes cease. Rather, equilibrium is maintained when the rate of each possible process is exactly balanced by the rate of the reverse process. The label 'detailed' in detailed balance is to remind us that this balance can be imposed at any level of specification that can be resolved, however detailed. An early powerful application of the 'detailed' part was Einstein's 1917 prediction of stimulated emission. This was based on the requirement that, at equilibrium, the rate of radiative transitions from a lower level i to an excited level f is exactly balanced by the rate of emission. The two detailed rates are to be separately balanced even though there can be other routes, e.g. collisional, for transferring population between these two levels.

Detailed balance is easy to state and discuss for thermal systems. The reason is that we all know and agree as to what are the concentrations of different species at thermal equilibrium. Then, for the forward and reversed elementary reactions

$$A + BC \underset{k}{\overset{k}{\rightleftarrows}} AB + C ,$$

and using square brackets to denote concentrations, with the subscript e denoting 'at equilibrium', detailed balance states that the forward and reversed rate constants are related by

$$\genfrac{}{}{0pt}{}{\text{forward reaction rate}}{\text{at equilibrium}} \equiv k[A]_e[BC]_e \equiv k[AB]_e[C]_e \equiv \genfrac{}{}{0pt}{}{\text{reverse reaction rate}}{\text{at equilibrium}} . \tag{8.3}$$

When the system is at equilibrium, the detailed balance condition, (8.3), is the condition that a dynamic state of equilibrium is maintained. If the system is not in equilibrium, the detailed balance condition, (8.3), is still satisfied. The difference is that, if the system is not in equilibrium, then the concentrations of the different species are not at their equilibrium values. Therefore, the left hand side of (8.3) is not the rate of the forward reaction. Rather, the forward rate is $k[\text{A}][\text{BC}]$ where the concentrations are at their instantaneous value. But (8.3) is still valid and, since we know the concentrations at equilibrium, it can be used to determine the ratio of the two rate constants. As a very familiar application, we determine the direction in which the reaction goes:

$$\frac{\text{forward reaction rate}}{\text{reverse reaction rate}} = \frac{k[\text{A}][\text{BC}]}{k[\text{AB}][\text{C}]} = \frac{[\text{A}][\text{BC}]/[\text{AB}][\text{C}]}{[\text{A}]_e[\text{BC}]_e/[\text{AB}]_e[\text{C}]_e}. \tag{8.4}$$

Therefore, if the mass ratio, $[\text{A}][\text{BC}]/[\text{AB}][\text{C}]$, exceeds the equilibrium constant the reaction moves forward and otherwise it moves backwards. A not quite so familiar result is obtained if we note that a 'molecule' is any species that can be experimentally distinguished. So that the chemical molecule AB in the quantum state j is a molecule in its own right, as is the chemical molecule BC in the quantum state i. A much more detailed form of (8.3) is therfore

$$k(i \to j)[\text{A}]_e[\text{BC}(i)]_e = k(j \to i)[\text{AB}(j)]_e[\text{C}]_e. \tag{8.5}$$

This is quite an explicit and useful relation because $[\text{BC}(i)]_e/[\text{BC}]_e$ is just the Boltzmann factor for the state i. Therefore (8.5), summed over i and j reads

$$\sum_i \sum_j \frac{[\text{BC}(i)]_e}{[\text{BC}]_e} k(i \to j)[\text{A}]_e[\text{BC}]_e = \sum_i \sum_j \frac{[\text{AB}(j)]_e}{[\text{AB}]_e} k(j \to j)[\text{AB}]_e[\text{C}]_e. \tag{8.6}$$

This is just (8.3). To see this one needs to remember the basic result [38] that the rate constant is to be summed over all final states but averaged over all initial states. Therefore, for the forward rate constant

$$k = \sum_i \sum_j \frac{[\text{BC}(i)]_e}{[\text{BC}]_e} k(i \to j), \tag{8.7}$$

and similarly for the reversed rate.

At this point one can clear up a misconception that some people have. It follows from the steps leading from (8.5) to (8.6) that the distribution of final states j in the forward reaction is whatever that distribution is, as specified by the dynamics. Each state j is produced with its dynamically determined detailed rate, $k(i \to j)$. There is no implication whatsoever that the distribution of final states j in the forward reaction is an equilibrium one. On the other hand, the distribution of initial states j in the reversed reaction must be an equilibrium one, if the averaged result, (8.3) is to be valid. Any initial state not specified explicitly by the nature of the species must have an equilibrium distribution, for detailed balance to be applied. A number of published applications of detailed balance are at fault because they overlooked the condition on the initial states. Yet the concerns that people tend to have are the unnecessary restriction on the final states.

The equation most relevant to our needs is intermediate between (8.5) and (8.6) and requires summation over only one index, say j:

$$\sum_j \frac{[BC(i)]_e}{[BC]_e} \boldsymbol{k}(i \to j)[A]_e[BC]_e = \sum_j \frac{[AB(j)]_e}{[AB]_e} \boldsymbol{k}(j \to i)[AB]_e[C]_e =$$

(8.8)

$$\boldsymbol{k}(i \to all)[A]_e[BC(i)]_e = \boldsymbol{k}(\to i)[AB]_e[C]_e \ .$$

For the unimolecular dissociation AB → A + B(i) this reads

$$\boldsymbol{k}(i \to all)[A]_e[B(i)]_e = \boldsymbol{k}(\to i)[AB]_e \ . \tag{8.9}$$

For our purpose, AB is a molecule in equilibrium while B is an electron with a given kinetic energy. Equation (8.9) related the rate constant for ionization into a given narrow kinetic energy range of the electron, $\boldsymbol{k}(\to i)$ to the rate of capture of an electron of a given kinetic energy by the molecular core A. Restrictions: As we already noted, in (8.6) all reactant quantum numbers not specified must have an equilibrium distribution. Not specified on the right of (8.9): The internal states of the molecule AB. The molecule (or cluster or whatever) AB must have an equilibrium distribution over all its internal quantum states. Otherwise (8.9) is not valid. Some authors have recently claimed that the application of detailed balance does not require an equilibrium assumption about the hot molecule, denoted AB in the above. This is not correct. Another point to note is that the quantum states of the core (denoted A) are not specified on the left of (8.9). The reaction rate is for a capture by A in internal equilibrium, not by A in its ground state.

Measuring electron capture by a core in its equilibrium state is not an easy task. How about specifying the state of the core for the capture process? No problem. Let the state of A be i'. Then (8.9) reads

$$\boldsymbol{k}(i \text{ and } i' \to all)[A(i')]_e[B(i)]_e = \boldsymbol{k}(\to i \text{ and } i')[AB]_e \ . \tag{8.10}$$

This equation relates the rate of capture of electrons of given kinetic energy by a specified core to the rate of ionization into a narrow range of kinetic energy of the electron *and* a core produced in a specified state. This result is not quite as useful as (8.9) because the rate of ionization that can be determined from this equation is the rate of ionization when the core is left in a defined state, the same state that does the capture in the reversed process. The application of detailed balance is very powerful but it does require attention. What we do not need to know is if the capture process, [A(i')] + [B(i)] does or does not produce AB molecules in equilibrium. Detailed balance does not ask about the distribution of unresolved final states. But detailed balance does insist on the 'detailed'. The named states on the left and right must be the same.

For the needs of our topic it is often more useful to consider equilibrium at a narrow range of the total energy. Everything goes exactly the same. The point that requires attention is simply that we are more familiar with computing concentrations for a system in thermal equlibrium. For a system at equilibrium at a narrow range of the total energy, a so-called 'microcanonical' equilibrium, one determines concentrations by counting states. This is because, at microcanonical equilibrium all quantum states are equally probable. The equilibrium 'concentration' of any species is therefore proportional to the number of quantum states that are implied by the nature of the species. That this really gives a concentration follows because each molecule has a center of mass and one must not forget to count the translational contribution. To make this explicit, let the kinetic energy of the center of mass be in the range $\varepsilon, \varepsilon + \delta\varepsilon$. The number of quantum states when the center of mass is confined to a box of length L and the momentum is in the range $p, p + \delta p$ is $(L\delta p/h)^3$ where the power 3 comes from the three spatial dimensions and V is the volume of the box. The number per unit volume of quantum states of the center of mass is $\delta \mathbf{p}/h^3$. Now $\delta \mathbf{p}/h^3 = 4\pi p^2 \delta p/h^3 = 4\pi mp \delta\varepsilon/h^3$. So, the number per unit volume of quantum states of the center of mass, when the kinetic energy is in the range $\varepsilon, \varepsilon + \delta\varepsilon$ is $\rho_T(\varepsilon)\delta\varepsilon = 4\pi mp\delta\varepsilon/h^3$. $\rho_T(\varepsilon)$ is known as the translational density of states. As is clear from the derivation, it is the density of states per unit volume. In addition to the translation one must also count the internal states. For example, a diatomic molecule with given electronic, vibrational and rotational quantum numbers has $2j+1$ internal quantum states, this being the number of distinct states with the same rotational quantum number j.

The total density of states corresponds to the situation when the total energy is in the range $E, E + \delta E$. This total can be partitioned in different ways between internal energy and the translational energy of the center of mass. In computing the total density of states one must sum over all possible such partitions. The summation is discrete because the internal energy is quantized. For example, for a diatomic molecule with given electronic and vibrational quantum numbers we need to compute the sum $\sum_{j=0}^{j_m}(2j+1)\rho_T(E - E_{\text{VIB}} - E_{\text{ROT}})$. Here E_{VIB} and E_{ROT} are the vibrational and rotational energy. The sum is finite and is terminated by the condition that no energy is available for the translation as the rotation took (almost) all, $E_{\text{ROT}} \leq E - E_{\text{VIB}}$. j_m is the quantum number of the highest allowed rotational state. The sum is easily approximately evaluated by assuming the molecule to be a rigid rotor. Then, for the total density of states

$$\rho(E) = \sum_{\nu=0}^{\nu_m} \sum_{j=0}^{j_m} (2j+1)\rho_T(E - E_{\text{VIB}} - E_{\text{ROT}}) , \qquad (8.11)$$

where ν_m is the vibrational quantum number of the maximal vibrational state allowed by conservation of energy, $E_{\text{VIB}} \leq E$. For a polyatomic molecule, there are more summations but these are readily carried out by a computer or approximately evaluated in closed form by assuming harmonic vibrations. Real difficulties do arise when higher lying isomers are energetically accessible, or, in general, when barriers to high amplitude internal motions can be surmounted. One knows what to do. For an n-atomic molecule in a given electronic state[1] it is necessary to evaluate (by Monte Carlo or otherwise) the $6n$ dimensional integral

$$\rho(E) = V^{-1} \int \delta(E-H)\,\mathrm{d}\mathbf{p}\,\mathrm{d}\mathbf{q}/h^{3n}\ . \tag{8.12}$$

Here $\mathbf{p}$ and $\mathbf{q}$ are the (vector) momenta and position of each one of the atoms. The real problem is not the evaluation of the integral. It is not easy, but when need be, it can be done. The real problem is to know the realistic potential when large amplitude motions are feasible.

Transition state theory is an example of a theory that satisfies the condition of detailed balance. To show this consider a reaction (i.e., crossing the barrier from left to right) and its reversed reaction crossing the barrier from right to left). The forward and reverse rate constants each have the functional form (8.2). The concentration of the species at microcanonical equilibrium is $\rho(E)\delta E$. Detailed balance is therefore the condition that, in either direction, the product $\rho(E)k(E)$ has the same value. But from (8.2), this is the case and the common value is the number, $N_\ddagger(E-E_0)$, of states of the system at the transition state. This number is the same in either direction because of the 'detailed' in detailed balance. If we are discussing a reaction and its detailed reversed reaction, the transition state must be the same. Despite the validity of detailed balance, transition state theory is not necessarily exact. Detailed balance is only a necessary condition. It is not, by itself a sufficient condition for a theory to be exact.

For the special case of delayed ionization one usually [39,40] applies detailed balance in the form of (8.9) with i being the kinetic energy ε of the outgoing electron. On the right, one has the rate of ionization into a narrow range in the final kinetic energy. This is balanced by the rate of capture, on the left hand side of the equation. Since the electron has a well defined velocity, the capture rate constant can be written as a product of the velocity times the capture cross section. We reiterate that this is the rate constant or, equivalently, cross section averaged over all the accessible states of the core. Both sides of the equation are to be evaluated at the same total energy, E. Say we take the zero of energy at the ground state of the unionized species. On this scale the ground state of the core will be at a higher energy, which is the threshold ionization energy, IP. At the given total energy E and when the electron takes the kinetic energy ε, the core is left with the energy $E-IP-\varepsilon$ above its ground state. In the simplest case of one isomer of the core (which also implies a given electronic level), the density of internal states of the core can be written as $g_c\rho_c(E-IP-\varepsilon)$ where g_c is the degeneracy of the electronic level of the core and ρ_c is the density of internal vibrational states. We emphasize the choice of the zero of energy by explicitly indicating the energy

---

[1] When several electronic states need to be included the problem is more complicated.

available to the core. The attachment rate is the rate averaged over all these states. We are unable to recommend the commonly made assumption that the capture rate is essentially independent of the internal state of the core. The internal density of states of the unpolarized electron is 2, the number of states of the spin. The translational density of states is, defacto, the translational density of states of the electron because the reduced mass for the relative motion of the electron and the core is the mass m_e of the electron. Detailed balance then gives for the total energy E the rate constant of emission (per unit energy of the outgoing electron) as

$$k(\to \varepsilon) = (16\pi m_e/h^3)\varepsilon\sigma_{\text{capture}}(\varepsilon)\frac{g_c\rho_c(E-IP-\varepsilon)}{g\rho(E)} . \tag{8.13}$$

To derive this result we used the density of translational states as obtained above and wrote the capture rate constant as $\nu\sigma_{\text{capture}}(\varepsilon)$ where $\varepsilon = m_e\nu^2/2$. To obtain the rate constant for electron emission we need to sum (= integrate) over the energy of the outgoing electron. At this point one needs the capture cross-section. The simplest model is to assume that electrons with angular momenta $< L$ are captured and others are not. Then $\sigma_{\text{capture}}(\varepsilon) = (\pi/k^2)L^2$ where $\varepsilon = \hbar^2k^2/2m_e$. Allowing L to be a function of ε, as it surely is, integration of (8.13) over ε yields

$$k(E) = \frac{1}{\hbar\rho(E)}\int_0^{E-IP} d\varepsilon(L(\varepsilon))^2\rho_c(E-IP-\varepsilon) . \tag{8.14}$$

This is just the result of transition state theory for a loose transition state. The number of states at the transition state is a convolution of the density of states for the rotation of the electron and that of the core.

The application of the statistical approach to delayed ionization is, in principle well understood. It has one quantitative problem, a problem that is serious enough to almost raise it to a point where it is a problem of principle: At the high energies under consideration, computing densities of states is not simple. There are two major obstacles. One is the vibrational density of states for a given electronic state of the system. A state counting based on harmonic oscillator frequencies is bound to be off by orders of magnitude, even if zero point energies and anharmonic corrections are applied. The reason is that the potential energy landscape is qualitatively not harmonic-like. There are many local minima that are connected by accessible bottlenecks. (For an attempt to allow for the different known higher energy isomers of C_{60} see [41]). Over and above this first difficulty there is the need to realistically account for the different electronic states that can take part. Already for a diatomic molecule, the density of electronic states of the neutral at such energies that ionization is possible, is enormous. For a metallic cluster, this is much more so. For the purpose of state counting, even a model of electrons in a box is more realistic than a conventional quantum chemical approach. These problems need to be addressed and they point out that the phase space of our system is anything but a simply connected smooth region. It is rough, convoluted and twisted with many bottlenecks that will slow the system on its way to sampling the energetically available phase space. But this means that one should be able to discern non-statistical effects. We turn to the experimental evidence.

8.4 Experimental: The Rate of Thermionic Emission

Long timescale (μs) delayed ionisation or "thermionic emission" has been studied extensively for fullerenes and metal clusters. A recent review has given a comprehensive coverage of the field [33]. Fullerenes are particularly interesting because of the strong competition with other cooling mechanisms such as neutral fragmentation and radiative cooling. Fullerenes are also interesting since they provide an intermediate between small metallic clusters and large molecules and show a number of similarities with both. The phenomenon was first observed in connection with the delayed production of ions after excitation with UV laser pulses [7,9]. A typical delayed ion tail,

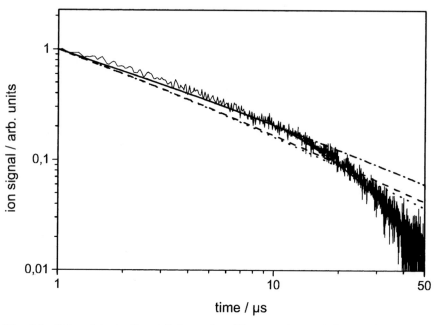

Fig. 8.1. Delayed ionisation tail from C_{60}. The experimental data is fitted using different combinations of competing processes [42]. *Dashed line*: only delayed ionisation. *Dots*: delayed ionisation and radiative cooling. *Dash-dot*: delayed ionisation, radiative cooling and high activation energy fragmentation. *Full line*: delayed ionisation, radiative cooling, high activation energy fragmentation with time-dependent contribution of low activation energy process. See text for more details. All fits have been normalised to the experimental data at a delay time of 1 μs

measured with 337 nm, 2 ns laser pulses is shown in Fig. 8.1 [42], plotted on a log-log scale. The arrival time of the ions after the prompt C_{60}^+ signal is a direct measure of the delay time for the electron emission. The signal clearly extends to times up to 50 µs and beyond. Earlier experiments using different laser wavelengths and fluences (ns pulse durations) showed essentially the same behaviour but over a more limited range [9]. Similar behaviour is seen when the electrons rather than the positive ions are detected. Echt and co-workers have carried out an extensive series of experiments for different laser fluences and temperatures of the oven used to produce the C_{60} beam [43]. A summary of their results is given in Fig. 8.2. The decay in all cases shows a power law behaviour (Intensity ($\propto t^{-n}$) rather than exponential due to the large range of internal energies produced in the experiments. A simple analysis by Hansen and Echt [39] where they assumed that fragmentation of the neutral C_{60} is competing with (and dominating over) the electron emission, shows that the slope of the plots, n, is given by the ratio of the ionisation potential (7.6 eV) to the dissociation energy. This analysis provided the first experimentally determined value of the dissociation energy for the neutral decay channel ($C_{60} \to C_{58} + C_2$) that did not rely on assumptions about the transition state. For all data measured by different groups and for different excitation wavelengths and frequencies, the slope, measured on a timescale up to 10 µs, is determined to be close to −0.7. This implies a large activation energy for the neutral dissociation channel on the order

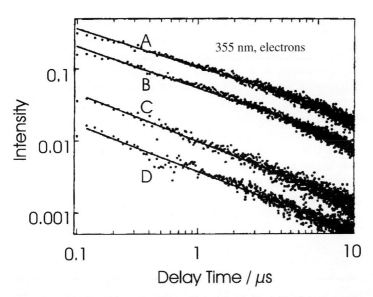

Fig. 8.2. Delayed electrons from C_{60} recorded with a laser wavelength of 355 nm and a range of source temperatures and laser fluences. *A:* 36 mJ/cm², 460 °C; *B:* 94 mJ/cm², 370 °C; *C:* 94 mJ/cm², 340 °C; *D:* 94 mJ/cm², 320 °C. *Full lines* represent least-squares fits of a power law. (adapted from [39])

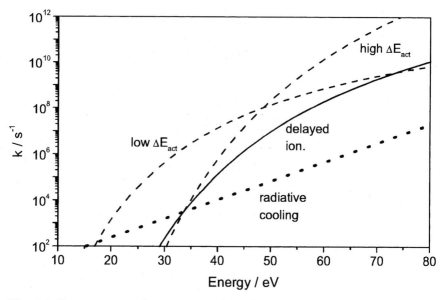

Fig. 8.3. Rate constants for competing decay processes in C_{60}. *Full line*: delayed ionisation [40]. *Dotted line*: radiative cooling [65]. *Dashed lines*: neutral fragmentation $C_{60} \rightarrow C_{58} + C_2$. High activation energy (10 eV) from [44]. Low activation energy (4.4 eV) from [46]

of 10–11 eV. Recently, such a high value with a correspondingly high pre-exponential factor in the Arrhenius-like rate constant for dissociation has found favour, also with the mass spectrometry community who extract the values from studying metastable decay fractions on a μs timescale [44]. This is in contrast with thermal experiments on a longer ms timescale that suggest a much lower activation energy and pre-exponential factor [45,46]. The data in Fig. 8.1 is compared to calculations in which the rate constants for delayed ionisation, neutral fragmentation and radiative cooling, shown in Fig. 8.3 have been used to model the time development of the positive ion production. For the fits shown in the figure, an initial broad, Poissonian internal energy distribution corresponding to the average absorption of 9 photons (33 eV) was assumed [42]. Delayed ionisation alone can clearly not reproduce the experimental data. A very good fit is obtained up to a time of ca. 10 μs by considering both ionisation and neutral fragmentation using the rate constant corresponding to a high activation energy. The slope obtained is in excellent agreement with experiment and the predictions of the Hansen-Echt analysis [39]. However, the long time behaviour can not be reproduced in this way. Both the electron measurements (Fig. 8.2) and the positive ion measurements (Fig. 8.1) show a faster drop in intensity than predicted. Taking radiative cooling of the neutral fullerenes into account makes a small difference but not sufficient to reproduce the experimental data[2]. Much better agreement is obtained however, if a second neutral dissociation channel is assumed to onset at 10 μs and gradually increase in importance as the time increases. In order to obtain the fit in the figure we have "switched on" the rate constant obtained by Kolodney and co-workers for thermal ms-timescale experiments [46]. This has a much lower activation energy and pre-exponential factor than the initial rate constant. The actual contribution of the second rate constant in Fig. 8.1 is still rather low (ca. 5% at 50 μs) but it is sufficient to reproduce the experimental data. An extrapolation to longer times would imply that the low activation energy rate constant should dominate on the ms timescale of the Kolodney experiment. This provides a very nice indication of the importance of a bottleneck in the distribution of energy and shows that the situation is considerably more complex than normally assumed.

8.5 Experimental: Dynamics

On excitation with ns laser pulses, there is enough time between the absorption of consecutive photons for the electronic excitation energy to be coupled to vibrational excitation. In this way the details of the initial excitation are erased and the system can be highly vibratonally excited at the end of the typically 10 ns long laser pulse. Recent experiments have shown how this situation changes in C_{60} as the timescale for the excitation is decreased [47,48]. Figure 8.4 shows extracts from time-of-flight mass spectra obtained with 790 nm wavelength laser light at a fluence of $5.5\,\mathrm{J/cm^2}$ [48]. The spectra have been obtained for different laser pulse durations, keeping everything else constant. The onset of the long timescale delayed ionisation can be seen to occur for pulse durations between 500 and 750 fs. Beyond this the behavior is very similar to that observed for excitation with ns laser pulses. The slope of the delayed tail measured on a log-log plot is in agreement with the slopes discussed in the previous section. As the pulse duration increases there is a longer time available for coupling to vibrational degrees of freedom, and the relative contribution of μs delayed ionisation compared to prompt ionisation increases. The threshold behavior is shown in Fig. 8.5 and is seen to be fluence dependent as one would expect. It is interesting to note in Fig. 8.4 that the internal vibrational energy of the fragment ions in this series of experiments, as indicated by the relative intensity of the broad metastable fragment peak to the prompt well-resolved mass peak, is not dependent on the pulse duration.

---

[2] Note added in proof: It was originally suggested that the observed behavior was a consequence of delayed ionization occurring mainly from the excited triplet state of C_{60} which decayed to the ground state in the time scale of tens of μs [42]. We now know [M. Hedén, A.V. Bulgakov, K. Mehlig, and E.E.B. Campbell (2003)] that the lifetime of the triplet state under the appropriate excitation conditions is too short to provide an explanation for the data. Here we suggest an alternative scenario.

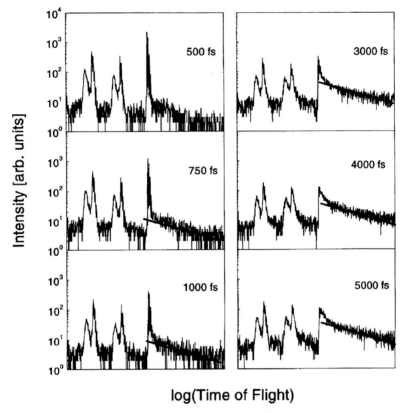

Fig. 8.4. Extract of time-of-flight mass spectra from C_{60} taken with identical laser fluence but different pulse durations. Note the onset of the delayed ionisation tail for durations between 500 fs and 750 fs and the increasing contribution of delayed ionisation as the pulse duration is increased. (from [48])

More detailed information can be obtained from photoelectron spectra [47]. The electron kinetic energy distribution measured for excitation with ps pulses is predominantly thermal as illustrated in Fig. 8.6c [47]. As the pulse duration is decreased below the threshold for µs delayed ionisation, the electron kinetic energy distribution remains predominantly thermal but increases very significantly in kinetic energy/temperature. This is clearly seen in the comparison between Fig. 8.6c and b where the laser fluence i.e. the total energy available in the laser pulse for exciting the molecule, remains constant. So, for pulse durations of ca. 500 fs and below the electrons that are emitted are still thermal, but the excitation energy has only been statistically distributed among the electronic degrees of freedom. The emission, although thermal, is "prompt" on the microsecond timescale of mass spectrometers and the parent mass peak is thus well-resolved with no tail towards longer flight times. On this timescale it is not possible to transfer electronic excitation

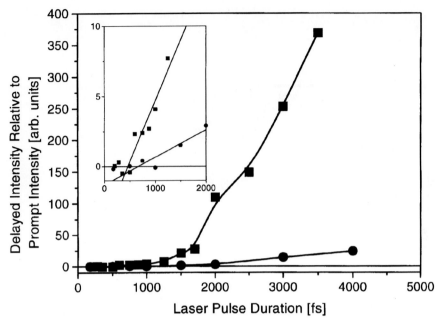

Fig. 8.5. Onset of delayed ionisation from C_{60} as a function of pulse duration for two different laser fluences. *Squares*: 5.5 Jcm^2, *Circles*: 4.6 Jcm^2. The lines in the main diagram are to aid the eye. The lines in the insert are least squares fits used to determine the onset of delayed ionisation. (from [48])

energy to vibrational degrees of freedom during the laser pulse. This means that the high electronic excitation produced on a very short timescale can lead to the production of multiply charged ions [49,50] as well as a rapid heating of the electronic subsystem of states which then thermally emits an electron before equilibration with the vibrational modes can take place. The same mechanism for electron emission is seen to occur in sodium cluster cations excited with fs laser pulses [51]. Such clusters do not undergo a µs delayed ionisaton since the dissociation energy is much lower than the energy needed to remove the second electron and fragmentation is thus the only observable decay channel on the µs timescale. Figure 8.7 shows the electron kinetic energy spectra from Na_{93}^+ as a function of laser fluence for excitation with 200 fs laser pulses. The high temperature in the electronic manifold of states in this case comes from the rapid decay of the plasmon resonance [52] which is excited resonantly by the 400 nm photons [51].

When the laser pulse duration is decreased even further – below 50 fs – the situation with C_{60} changes again. Experiments with 25 fs laser pulses show no thermal electrons for the same fluence as in Fig. 8.6b and c, but instead show very clear above threshold ionisation peaks [47]. For this very short timescale there is not even enough time for the electronic energy to be

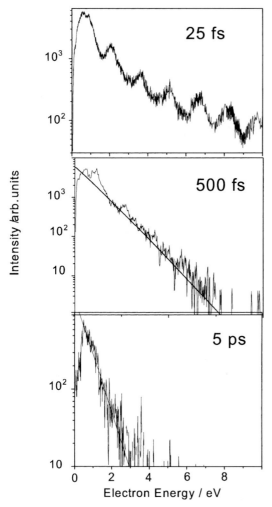

Fig. 8.6. Photoelectron spectra obtained for approximately the same laser fluence (3 J/cm^2, 790 nm) but different pulse durations illustrating the different ionisation regimes observable in C$_{60}$. (**a**) 25 fs; above threshold ionisation (**b**) 500 fs: prompt thermal emission of electrons from electronic sub-system (**c**) thermal electron emission after coupling to vibrational degrees of freedom "thermionic emission". (adapted from [47])

statistically distributed among the electronic degrees of freedom during the laser pulse leading to a strong contribution of multiply charged ions in the mass spectrum [50]. The ionisation is seen to be predominantly sequential multiphoton ionisation producing parent ions up to C$_{60}^{5+}$.

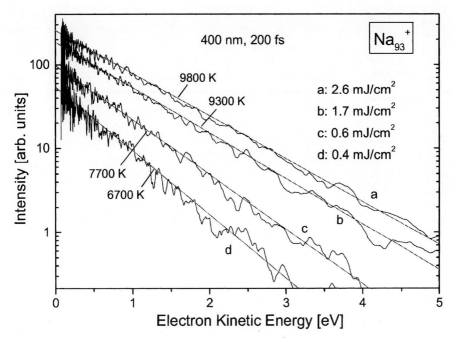

Fig. 8.7. Photoelectron spectra obtained from Na_{93}^+ with 200 fs pulses at 400 nm and for different laser fluences. The emission is predominantly thermal in origin and comes from the very hot electron bath before coupling to vibrational modes occurs (similar to Fig. 8.6b). (adapted from [51])

8.6 Kinetic Model

The quantum theory of decay processes with more than one time scale is well developed, [34–36]. In essence the key consideration is a competition between decay rates from different regions of phase space and intramolecular mixing processes which tend to equalize the population and thereby erase any non statistical behavior. In the limit where the intramolecular mixing is fast, the decay is rate determining and one has the limit envisaged in transition state theory. Deviations from the statistical limit are possible when different regions of phase space communicate more slowly; when bottlenecks to intramolecular dynamics are as important as the bottlenecks to the decay. This physical aspect can be recovered in a simple model, Fig. 8.8, when phase space is broken into distinct but connected regions, each with its own decay rate. The simplest such scheme is shown in Fig. 8.8 where only the two regions, that are experimentally slowest to come to equilibrium, are drawn. The slow interconversion is depicted as a rate process.

The smaller region in Fig. 8.8 is where the energy is localized in the electronic degrees of freedom, so that the mean energy is high. Electron emission can still be delayed because no particular electron has enough energy

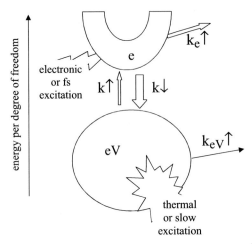

Fig. 8.8. A schematic representation of a phase space with an internal bottleneck between the manifold of electronically excited states, labeled e, and the rest of phase space, eV, where the excitation energy is distributed over both electronic and vibrational modes. The rate constants for intramolecular communication are denoted $k \uparrow$ and $k \downarrow$ where the direction of the arrows denotes the change in energy per degree of freedom. The rate constant for ionization from the e manifold, $k_e \uparrow$ is higher than that for quasiequilibrium, $k_{eV} \uparrow$ because of the higher mean velocity of the electrons

to escape promptly. The bigger chunk of phase space is the one where the vibrations also take up some of the energy. It is a bigger region and so, by detailed balance, one expects that it is easier to get into this region, (rate constant $k \downarrow$), than to get out, (rate constant $k \uparrow$). If the initial preparation is into the bigger region then the presence of the smaller region will be hardly noticed. One can write down kinetic rate equations for the loss of molecules from the two regions, e.g.

$$-\frac{d[eV]}{dt} = (k_{eV}\uparrow + k\uparrow)[eV] - k\downarrow[eV]. \tag{8.15}$$

These kinetic equations can easily be analytically solved but it is intuitively clear that unless $k \uparrow$ is large, the system will not sample the pure electronic, (denoted 'e' in Fig. 8.8), region, as shown graphically in Fig. 8.9.

On the other hand, if initially the excitation is primarily into the electronic manifold, there will be two slow decay modes, a faster one due to delayed ionization from the e manifold of states, and the usual slow one, Fig. 8.9

$$-\frac{d[eV]}{dt} = (k_{eV}\uparrow + k\downarrow)[eV] - k\uparrow[eV]. \tag{8.16}$$

The amplitude of the two decay modes depends on the competition between the intramolecular mixing and the rates of electron emission. For longer times

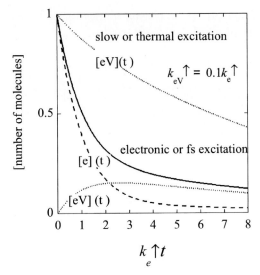

Fig. 8.9. A graphical representation of the time evolution of the populations in phase space for two different modes of initial excitation. Computed for the same set of rate constants with $k_e\uparrow = 10\,k_{eV}\uparrow$. Realistically, the ratio of the two ionization rate constants should be higher but then the two cases cannot be shown in the same plot against time. For a non-selective excitation there is hardly any population in the e manifold. For initial excitation in the electronic manifold there is a faster initial decay, where the population mostly stays in the e manifold. For longer times, the two manifolds equilibrate and the much slower decay of the total population, (*solid curve*), is effectively through the eV manifold

the internal distribution stabilizes as can be seen in the bottom part of Fig. 8.9, where the population in eV, initially zero, rises.

8.7 Concluding Remarks

High molecular Rydberg states [29,30,53–60] provide the clearest example of delayed ionization that is not thermionic. The physics of this delay is well understood. The Rydberg electron has a two way protection from interaction with the molecular core, an interaction which can provide the tiny excitation energy required for its departure. First, the Bohr-Sommerfeld orbit for high n but low ℓ states only gets near the core once per orbit and the orbital period is rather long. Beyond that, by increasing the initially low value of ℓ one can further stabilize the electron. Time resolved experiments [29,61,62] have clearly established decays on the sub-µs time scales while at shorter times there is both prompt ionization and ionization delayed on the ns time scale [63,64]. The experiments discussed in this chapter extend such observations to even shorter times. In the face of all the available experimental and theoretical evidence it seems appropriate to conclude that delayed ionization proceeds through several stages only the last of which is fully statistical.

Acknowledgements. Our work owes a lot to Uzi Even, Ingolf Hertel, Françoise Remacle, Frank Rohmund and Edward Schlag, SFB 377, SFB 450 and the European Network on the Cooling of Clusters (HPRN-CT-2000-00026). The work on extreme conditions is supported by the Humboldt Foundation.

References

1. S. Dushman, Rev. Mod. Phys. **2**, 381 (1930).
2. R.H. Fowler, Statistical Mechanics (Cambridge University Press, Cambridge, 1936).
3. C. Kittel, Introduction to Solid State Physics (John Wiley & Sons, Inc., New York, 1953).
4. P.D. Dao and J.A.W. Castleman, J. Chem. Phys. **84**, 1435 (1986).
5. A. Amrein, R. Simpson and P. Hackett, J. Chem. Phys. **94**, 4663-4664 (1991).
6. A. Amrein, R. Simpson and P. Hackett, J. Chem. Phys. **95**, 1781 (1991).
7. E.E.B. Campbell, G. Ulmer and I.V. Hertel, Phys. Rev. Letters **67**, 1986 (1991).
8. T. Leisner, K. Athanassenas, O. Echt, O. Kandler, D. Kreisle and E. Recknagel, Z. Phys. D **20**, 127 (1991).
9. P. Wurz and K.R. Lykke, J. Chem. Phys. **95**, 7008 (1991).
10. K. Athanassenas, T. Leisner, U. Frenzel and D. Kreisle, Ber. Bunsenges. Phys. Chem. **96**, 1192 (1992).
11. E.E.B. Campbell, G. Ulmer and I.V. Hertel, Z. Phys. D **24**, 81 (1992).
12. P. Wurz and K.R. Lykke, J. Phys. Chem. **96**, 10129 (1992).
13. K. Athanassenas, T. Leisner, U. Frenzel, D. Kreisle and E. Recknagel, Z. Phys. D **26**, 153 (1993).
14. B.A. Collings, A.H. Amrein, D.M. Rayner and P.A. Hackett, J. Chem. Phys. **99**, 4174 (1993).
15. T. Leisner, K. Athanassenas, D. Kreisle, E. Recknagel and O. Echt, J. Chem. Phys. **99**, 9670 (1993).
16. G. Walder, K.W. Kennedy and O. Echt, Z. Phys. D **26**, 288 (1993).
17. B.D. May, S.F. Cartier and J.A.W. Castleman, Chem. Phys. Letters **242**, 265 (1995).
18. R.D. Beck, P. Weis, J. Rockenberger and M.M. Kappes, Surf. Rev. Letters **3**, 771 (1996).
19. S.F. Cartier, B.D. May and J.A.W. Castleman, J. Chem. Phys, **104**, 3423 (1996).
20. G. Ganteför, W. Eberhardt, H. Weidele, D. Kreisle and E. Recknagel, Phys. Rev. Letters **77**, 4524 (1996).
21. H. Weidele, S. Becker, H.-J. Kluge, M. Lindinger, L. Schweikhard, C. Walther, J. Ziegler and D. Kreisle, Surf. Rev. and Lett. **3**, 541 (1996).
22. A. Bekkerman, B. Tsipinyuk, A. Budrevich and E. Kolodney, J. Chem. Phys. **108**, 5165 (1998).
23. S.E. Kooi and J.A.W. Castleman, J. Chem. Phys. **108**, 8864 (1998).

24. M. Quack, in Encyclopedia of Computational Chemistry, Vol. 3, edited by N.L. Allinger, P.v.R. Schleyer, T. Clark, J. Gasteiger, P.A. Kollmann, H.F. Schaefer III, P.R. Schreiner (John Wiley & Sons, Chichester, UK, 1998), pp. 1775–1791.
25. H. Weidele, D. Kreisle, E. Recknagel, S. Becker, H.-J. Kluge, M. Lindinger, L. Schweikhard, C. Walther and J. Ziegler, J. Chem. Phys. **110**, 8754 (1999).
26. S.T. Arnold, R.A. Morris and A.A. Viggiano, J. Chem. Phys. **103**, 9242 (1995).
27. C. Yeretzian, K. Hansen and R.L. Whetten, Science **260**, 652 (1993).
28. W.G. Scherzer, H.L. Selzle, E.W. Schlag and R.D. Levine, Phys. Rev. Letts. **72**, 1435 (1994).
29. U. Even, Phil. Trans. Roy. Soc. **355**, 1539 (1997).
30. E.W. Schlag and R.D. Levine, Comments At. Mol. Phys. **33**, 159 (1997).
31. G. v. Helden, I. Holleman, A.J.A. v. Roij, G.M.H. Knippels, A.F.G. v. d. Meer and G. Meijer, Phys. Rev. Letters **81**, 1825 (1998).
32. J.M. Weber, K. Hansen, M.-W. Ruf and H. Hotop, Chem. Phys. **239**, 271 (1998).
33. E.E.B. Campbell and R.D. Levine, Ann. Rev. Phys. Chem. **51**, 65 (2000).
34. F. Remacle and R.D. Levine, J. Chinese Chem. Soc. **42**, 381 (1995).
35. F. Remacle and R.D. Levine, in Dynamics of Molecules and Chemical Reactions, edited by R.E. Wyatt and J.Z.H. Zhang (Marcel Dekker, New York, 1996), pp. 1–59.
36. R.D. Levine, Adv. Chem. Phys. **101**, 625 (1997).
37. L.Y. Baranov, R. Kris, R.D. Levine and U. Even, J. Chem. Phys. **100**, 186 (1994).
38. R.D. Levine and R.B. Bernstein, Molecular Reaction Dynamics and Chemical Reactivity (Oxford University Press, New York, 1987).
39. K. Hansen and O. Echt, Phys. Rev. Letters **78**, 2337 (1997).
40. C.E. Klots, Chem. Phys. Letters **186**, 73 (1991).
41. E.E.B. Campbell, T. Raz and R.D. Levine, Chem. Phys. Lett. **253**, 261 (1996).
42. F. Rohmund, M. Heden, A. Bulgakov, E.E.B. Campbell, J. Chem. Phys. **115**, 3068 (2001).
43. R. Deng and O. Echt, J. Phys. Chem. **102**, 2533 (1998).
44. C. Lifshitz, Int. J. Mass. Spec. **198**, 1 (2000).
45. E. Kolodney, B. Tsipinyuk and A. Budrevich, J. Chem. Phys. **100**, 8542 (1994).
46. E. Kolodney, B. Tsipinyuk and A. Budrevich, Phys. Rev. Lett. **74**, 510 (1995).
47. E.E.B. Campbell, K. Hansen, K. Hoffmann, G. Korn, M. Tchaplyguine, M. Wittmann and I.V. Hertel, Phys. Rev. Lett. **84**, 2128 (2000).
48. E.E.B. Campbell, K. Hoffmann and I.V. Hertel, Eur. Phys. J. D **16**, 345 (2001).
49. M. Tchaplyguine, K. Hoffmann, O. Dühr, H. Hohmann, G. Korn, H. Rottke, M. Wittmann, I.V. Hertel and E.E.B. Campbell, J. Chem. Phys. **112**, 2781 (1999).
50. E.E.B. Campbell, K. Hoffmann, H. Rottke and I.V. Hertel, J. Chem. Phys. **114**, 1716 (2001).
51. B. von Issendorf and H. Haberland, Appl. Phys. A **72**, 255 (2001).
52. R.P. Schlipper, R. Kusche, C. Bréchignac and J.P. Connerade, J. Phys. B. **27**, 3795 (1994).
53. W.A. Chupka, J. Chem. Phys. **98**, 4520 (1993).
54. F. Remacle and R.D. Levine, Phys. Lett. A **173**, 284 (1993).
55. M. Bixon and J. Jortner, J. Phys. Chem. **99**, 7466 (1995).

56. F. Merkt, Annu. Rev. Phys. Chem. **48**, 675 (1997).
57. F. Remacle and R.D. Levine, J. Chem. Phys. **107**, 3382 (1997).
58. F. Remacle and R.D. Levine, Models in Chemistry ACH **134**, 619 (1997).
59. The Role of Rydberg States in Spectroscopy and Reactivity, Vol., edited by C. Sandorfy (Kluwer Academic, Dordrecht, 1999).
60. E.W. Schlag, *ZEKE Spectroscopy* (Cambridge University Press, Cambridge, 1998).
61. U. Even, M. Ben-Nun and R.D. Levine, Chem. Phys. Lett. **210**, 416 (1993).
62. U. Even, R.D. Levine and R. Bersohn, J. Phys. Chem. **98**, 3472 (1994).
63. F. Merkt and R.N. Zare, J. Chem. Phys. **101**, 3495 (1994).
64. F. Merkt, S.R. Mackenzie and T.P. Softley, J. Chem. Phys. **103**, 4509 (1995).
65. C. Lifshitz, Int. J. Mass. Spec. **198**, 1 (2000).

9 Cluster Dynamics: Influences of Solvation and Aggregation

Q. Zhong and A.W. Castleman, Jr.

9.1 Introduction

The vast majority of chemical reactions of practical interest take place in the condensed phase, but an understanding of the details of reaction dynamics has emerged largely from investigations in the isolated gas phase. One valuable approach to bridging an understanding of reactive behavior between the two phases is made possible through investigations of clusters [1]. The entire course of a chemical reaction following either a photophysical or ionizing event, depends on the mechanisms of energy transfer and dissipation away from the primary site of absorption [2–4]. The presence of neighboring solvent or solute molecules not only perturbs the energetics of the reaction coordinate, they can also influence the operative mechanisms through collisional de-activation and caging effects. Through the use of supersonic molecular beams, it is now possible to produce and tailor the composition of virtually any complex system of interest [5–8]. Therefore, one can selectively solvate a given chromophore (site of photon absorption) and investigate changes between the gas and the condensed phase by selectively shifting the degree of solvent aggregation, i.e., the number of solvent molecules attached to or bound about the site of photo-absorption. The vast majority of spectroscopically related cluster work is accomplished using clusters formed via supersonic expansion and detected using time-of-flight mass spectrometry techniques (TOFMS) [9], with the improved mass resolution from the incorporation of reflection electric fields (reflectron) [10–12]. The study of cluster dynamics by employing time resolved spectroscopy is a comparatively new subject of inquiry. Its potential has been brought to fruition through recent developments in picosecond and especially femtosecond laser pump-probe techniques [8,13]. By exploiting advances made in generating laser pulses of sub-picosecond duration, the field of femtochemistry was born. Through the pioneering and innovative concepts of the Zewail group [14] in implementing the pump-probe technique, many breakthroughs in this emerging field already have been witnessed over the past decade [13].

This chapter reviews some of the advances that have emerged from investigations of cluster systems using pump-probe spectroscopy. Our focus is first directed toward several characteristic mechanisms, namely ones involving photo-induced electron transfer, followed by a consideration of excited state single and double proton transfer. Next the role of caging dynamics is discussed, with attention to both neutral and anionic cluster systems. During the course of our studies of cluster dynamics using intense femtosecond laser pulses, an important phenomenon was uncovered in our laboratory, namely the influence of molecular clusters on the production of species of high charge state which subsequently lead to Coulomb explosion. We discuss some of the initial findings in this area, followed by work that led to the use of Coulomb explosion as a method to arrest reaction intermediates in fast reactions and follow the course of a chemical reaction [15–17]. The main emphasis of discussion in the afore mentioned areas is on hydrogen bonded and van der Waals clusters. We complete this discussion of cluster dynamics with an overview of recent findings on the excitation/electronic dynamics for the case of a unique semiconductor cluster system [18–64] discovered in our laboratory, which comprises of early transition metals bound to carbon in a specific composition ratio, namely 8:12. Study of the dynamics of these clusters [58–60], termed metallocarbohedrenes, or Met-Cars for short, allows insight into electronic relaxation into electronic bands, ways of obtaining information on the vibrational modes of the clusters, and further understanding of microscopic processes that can lead to the phenomenon of thermionic emission in certain cluster systems [56,57,61].

9.2 Charge-Transfer Reactions

Charge-transfer reactions are of fundamental importance in a wide range of chemical and biological processes, and there is a continuous and growing interest in determining the molecular details of the elementary charge-transfer steps [65–68] Since charge-transfer processes in natural environments are quite complex, many researchers have studied charge transfer reaction in well-defined cluster systems with well-characterized optical pulses as a route to understand the elementary aspects of this process free from complicating or competing processes found in natural environments. In this section, we discuss selected studies of photoinduced electron-transfer and proton-transfer processes.

9.2.1 Photo-Induced Electron-Transfer Reactions

When Benesi and Hildebrand [69] first dissolved iodine and benzene in n-heptane, they observed a new absorption band in the ultraviolet range. Mulliken [70] attributed this band to an electronic transition from the highest occupied molecular orbital of benzene to the lowest unoccupied molecular orbital of iodine. Since the initial discovery, extensive research efforts have been devoted to the investigation of this particular type of electron-transfer process in a variety of environments, including isolated bimolecular complexes [71,72] cluster systems [73–76], rare gas matrices [77], and liquid solutions [78–82].

The first time-resolved experiments on Bz•I_2 (Bz, benzene) were performed by Zewail and coworkers [71]. In their experiments, a femtosecond pump pulse at 277 nm launches the Bz•I_2 complex to its ionic potential energy surface (PES); the fragmented I atom was probed at 304 nm by kinetic energy resolved resonance multiphoton ionization (REMPI) mass spectrometry. In Bz•I_2, the bound ionic PES crosses the repulsive covalent PES and the avoided crossing of these two PESs forms an upper quasibound potential well and a lower potential well with a barrier at the crossing, as sketched in Fig. 9.1. Following photoexcitation to the ionic PES, the Bz^+I^-••I complex propagates along the ionic curve towards the hybrid region of covalent and ionic potentials. At the crossing barrier, some of the wavepacket population continues on the ionic PES, whereas the rest of the population crosses to the neutral PES through an electron back transfer to the Bz molecule, leaving the I–I bond on its dissociative potential. The release of the first iodine atom with high translational energy was observed in 450 fs. The observation of the release of the other "caged" iodine atom occurs in 1.4 ps, with a smaller translational energy compared to the first iodine atom due to collisional energy relaxation with the Bz molecule [71]. Further experiments involving different donor molecules like diethyl-sulfide, o-xylene, acetone, and dioxane reveal that the escape time of the second iodine atom from the single molecule cage increases when the interaction between the donor molecule and the I atom is stronger [71,72].

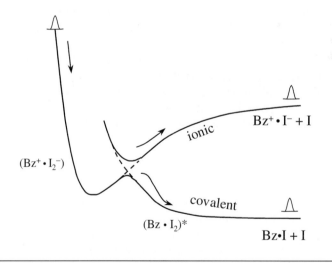

Fig. 9.1. A schematic of the potential energy curves of the dissociative electron transfer reaction along the reaction coordinate in Bz•I_2 system. (adapted from [71])

When the electron donor is solvated in a cluster, the electron-transfer rate is found to be strongly dependent on the degree of solvation [71,75]. In time-resolved experiments on $(Bz)_n I_2$ clusters, the releasing of the iodine atoms was also found to occur in two different time scales [71], as was observed in the bimolecular complexes discussed above. Theoretical calculations [75] using *ab-initio*, Monte Carlo, and molecular dynamics simulations indicate that this may be attributable to the structural asymmetry of the $(Bz)_n I_2$ clusters. The benzene molecules were found to cluster around themselves rather than to the iodine molecule, resulting in the iodine molecule protruding out of the benzene cluster, with one of the iodine atoms more strongly solvated than the other. Following the photoinduced electron-transfer resulting in the I–I bond rupture, the less solvated iodine atom leaves the cluster quickly, dragging some benzene molecules with it. This free iodine fragment was detected in sub-picosecond to picosecond time scale. The escape of the residual caged iodine atom occurs on a much longer time scale ($> 10\,\text{ps}$), and shows a time dependence on the vibrational energy of the cluster. The dissociation times of both iodine atoms increase with increasing degrees of solvation [71,75].

9.2.2 Excited-State Proton-Transfer

Heats of deprotonation of many molecules (e.g. aromatic alcohols) shows a dependence on electronic excitation, which leads to pH changes upon photoexcitation. For example, the pKa of 1-naphthol changes from 9.4 in its ground electronic state to 0.5 in its $S1$ state [68]. This dramatic decrease in pKa enables the proton-transfer process to be initialized with one laser pulse and the progress of the reaction can be followed by a second laser pulse, as demonstrated in various ultrafast pump-probe experiments in solution [68,83–85], solid matrices [86], and clusters [87–94].

Experiments in gas phase cluster systems [87–94] reveal that excited-state proton-transfer (ESPT) rates display a strong dependence on cluster size and the corresponding solvent structure. Since a relatively large amount of energy is needed to free a proton from the solute molecule, the ion-pair resulting from proton transfer is unstable compared to the original covalent structure. Solvation of the proton with a high proton affinity solvent significantly enhances the stability of the ion-pair. It is generally observed that the larger the solvent clusters, the larger the gained stability. In the case of $C_6H_5OH\bullet(NH_3)_n$ clusters [87,89,93], the proton-transfer time is measured to be 5 ns for $n = 4$, but decreases dramatically to 55 ps at $n = 5$. The dramatic rate change at $n = 5$ was interpreted as the signature of the onset of the ESPT reaction. In the case of 1–NpOH$\bullet(NH_3)_n$ clusters (NpOH, 1-naphthol, is more acidic in its S_1 state than phenol [pKa $= 4.1$]), ESPT was observed to occur at $n = 3$ [88,90] or $n = 4$ [91]. For solvents with smaller proton affinities such as water, much larger clusters are necessary before ESPT can be observed. Experiments [92] on 1–NpOH$\bullet(H_2O)_n$ show that ESPT occurs with a minimum water cluster size of $n = 20 \sim 30$. In experiments involving mixed solvents, the incorporation of different solvent molecules can vary the hydrogen-bonding network in the ion-pair structure, thus affecting the proton-transfer process dramatically. For example, the ESPT in photoexcited $C_6H_5OH\bullet(NH_3)_5$ occurs within 55 ps, however when one methanol molecule is added to the clusters, the proton-transfer time increases to 750 ps [89].

The experimentally observed excited state transient displays bi-exponential decay characteristics. For example, the 1–NpOH•$(NH_3)_5$ transient has a distinct double exponential decay, with a fast component of 25 ps and a slow component of 226 ps. The observed fast component was attributed to the proton-transfer process, while the slow component was interpreted to be due to solvent reorganization following the proton-transfer [89,90]. Solvent reorganization involves the rearrangement of the hydrogen-bonded network following the proton-transfer reaction, leading to a more stable ion-pair structure. The solvent reorganization rate shows dependence on cluster size, vibrational excitation, and also an isotope effect [90].

When deuterated species, for example, $C_6H_5OD\bullet(ND_3)_5$, is used, the ESPT rate slows down to 1.5 ns from 55 ps of the $C_6H_5OH\bullet(NH_3)_5$ [89]. This isotope effect was also observed in other systems [90]. The large isotope effect supports a tunneling mechanism. The tunneling barrier is formed by the avoided curve-crossing of the covalent potential with the solvent stabilized ion-pair potential. Experimental evidence for the existence of this tunneling barrier is the observed slow proton-transfer rate even in highly acidic 1-naphthol with high proton affinity solvents, like ammonia or trimethylamine. The solvent type, structure, and degrees of solvation all affect the potential energy surfaces and thus the barrier shape, therefore influencing the proton-transfer rate profoundly.

Some recent experimental results on $C_6H_5OH\bullet(NH_3)n$ clusters [95] have shown that for small clusters ($n = 1-5$), a possible previously overlooked hydrogen atom transfer reaction also plays an important role in the excited state dissociation process. The excited state hydrogen atom transfer (ESHT) occurs when photoexcited C_6H_5OH dissociatively donates a hydrogen atom to the solvating ammonia cluster. The resulting solvated ammonia radical has a microsecond scale lifetime and can be detected easily. In light of these findings, additional experiments and further analysis of previously obtained data taking into consideration the possible new channels are called for to further understand the excitation and dissociation mechanisms of these systems.

Competition between ESPT and ESHT channels has been observed in a pump-probe study of the Rydberg $\tilde{C}$ state of ammonia clusters in our lab [96]. We found that the $\tilde{C}$ state lifetimes of ammonia clusters are more than one magnitude shorter than that of the ammonia monomer. This is the opposite of the observed solvation effect for the predissociative $\tilde{A}$ state [97,98]. Dissociation of the ammonia monomer on its $\tilde{C}$ state occurs through coupling to the lower lying A state, followed by subsequent predissociation. As shown schematically in Fig. 9.2, two reaction channels are open to the

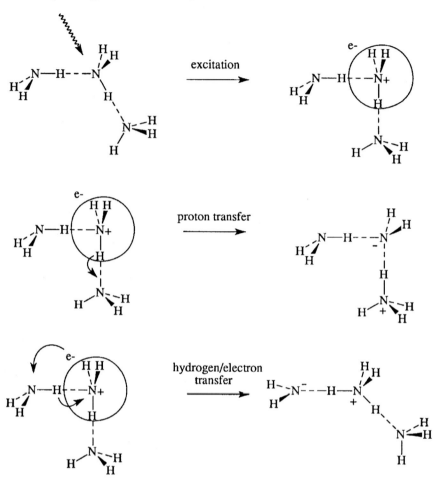

Fig. 9.2. Schematics of the proposed excited state proton and hydrogen transfer processes in the $\tilde{C}'$ state of ammonia clusters. (adapted from [96])

clusters, namely ESPT and ESHT. ESPT occurs when the photoexcited ammonia donates a proton to a solvating ammonia molecule, while ESHT occurs when the photoexcited ammonia accepts a hydrogen atom from a solvating ammonia molecule. These two reaction channels nicely explain the experimentally observed bi-exponential decay of the cluster transients [96]. Both reaction channels are dependent on the degree of solvation; however, the ESHT rate shows a stronger solvation dependence than ESPT rate. For example, the ESPT time for $(NH_3)_n$ increases from 85 fs for $n = 3$, to 135 fs for $n = 40$; while ESHT times increase from 0.3 ps for $(NH_3)_3$, to 1.5 ps for $(NH_3)_{40}$. Two competing factors attributable to a solvation effect may influence the observed reaction dynamics: 1) energy relaxation due to solvation; 2) solvation stabilization of potential energy surfaces. When ground state ammonia is excited to its $\tilde{C}$ state, the out of plane "umbrella" bend is highly excited. The "umbrella" vibration moves along the hydrogen atom transfer coordinate, therefore, excitation of the umbrella mode is expected to enhance the hydrogen atom transfer rate. As the cluster size increases, the energy relaxation in this umbrella mode becomes faster; thus the observed ESHT rate slows down [96]. Since the proton transfer reaction coordinate lies perpendicular to the umbrella vibration, it shows a smaller dependence upon clustering. Related ESHT in the ammonia dimer on the Rydberg $\tilde{A}$ state [99] also has been observed by Hertel's group, using femtosecond time-resolved photoelectron-photoion coincidence spectroscopy [100].

9.2.3 Excited-State Double Proton-Transfer

The 7-Azaindole dimer, a model system for understanding base-pairs such as DNA, shows a particularly interesting behavior when excited to its S_1 state – double proton transfer. This excited-state double proton-transfer (ESDPT) was first observed by Kasha et al. in solution [101], and thereafter numerous studies have been undertaken by various research groups in both the gas [15,17,102–106] and condensed phase [107–109]. A major disagreement developed in the literature over whether ESDPT proceeds through a step-wise or a concerted process. Gas phase experiments [15,106] generally supported a step-wise proton transfer mechanism, while condensed phase studies made in the presence of a nonpolar liquid showed either a concerted [107,108] or a step-wise [109] mechanism depending on the excitation energy.

The first time-resolved measurements on the 7-Azaindole dimer by the Zewail group revealed a two-step process for the ESDPT, with 650 fs and 3.3 ps for the first and second proton transfer times, respectively [106]. Our group arrived at the same conclusion utilizing a Coulomb explosion imaging technique (discussed in Sect. 9.4.) that can be used to arrest and detect evolving reaction intermediates. These new experiments [15,17] provided definitive evidence that a reaction intermediate exists, as suggested by Zewail, confirming the two-step process for ESDPT in the isolated dimer.

Studies of the 7-azaindole dimer by Kaya and co-workers [103] suggest the existence of two isomers. Depending upon the stagnation pressure employed in the supersonic expansion, the formation of one isomer is favored over the other. These isomers behave differently; one undergoes normal ESDPT (reactive), and the other not (non-reactive). They suggested that the structural difference of the two isomers were responsible for the observed difference in reactivity. Time-resolved measurements in our lab16 confirmed the existence of reactive and non-reactive dimers. The reactive dimer transient displayed a bi-exponential decay with time constants nearly identical to those obtained by Zewail [106]. The non-reactive dimer transient is stable, which indicated that the ESDPT had not occurred in the observation window of the experiments. A particularly interesting observation is that when the 7-azaindole

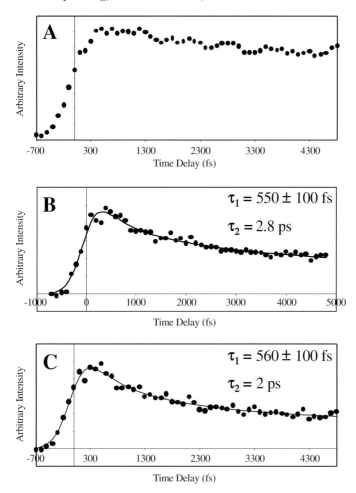

Fig. 9.3. The solvation effect of water clusters on the ESDPT process of 7-Azindole dimer. Adapted from [16]. (**a**) Pump-probe transient of the nonreactive dimer (stagnation pressure, 1 200 torr). (**b**) Pump-probe transient of the nonreactive dimer with one water clustered to it. It can be seen here that an attached water molecule actually facilitates the occurrence of the proton transfer in an otherwise nonreactive species. (**c**) Pump-probe transient of the nonreactive dimer with two waters clustered to it. The second step of the proton transfer can be seen to occur even more quickly with two waters present

dimer is solvated with water molecules, the reactivity changes completely – the non-reactive dimer becomes reactive. Figure 9.3 shows the non-reactive dimer transients with differing numbers of solvated water molecules. It can be seen from the figure that the non-reactive dimer becomes reactive even with only one solvating water molecule. The most striking observation is that subsequent solvation beyond one water molecule has little effect on the proton

Table 9.1. Proton transfer times for water solvated 7-azindole dimer, adapted from [16]

Number of Waters	First Transfer, fs	Second Transfer, fs	Transfer Time, fs
1	550 ± 100	2800	
2	565 ± 100	2000	
3	560 ± 100	2000	
4	– (a)	– (a)	
5			1800
6			1600
7			1300
8			1100
9			1000

(a) – no well established could be determined

transfer until a dramatic change occurs at four waters and above, where only a single proton transfer step is observed. Table 9.1 shows the proton transfer times at different degrees of water solvation. The results suggest that the double proton transfer is undergoing a transition from a step-wise to a concerted process. It is possible that this change is due to progressive clustering, but, alternatively, the proton may be transferred to the waters when the excited species have higher degrees of solvation. A preliminary calculation by Hobza [110] suggests that the 7-azaindole dimer can adopt a non-hydrogen-bonded stacked structure instead of the hydrogen-bonded structure once the cluster contains enough water molecules. It is believed that only one of the 7-azaindole monomer units is photo-excited in the stacked dimer [111], and therefore, the single exponential decay observed could result from a self-tautomerization by means of a water molecule acting as a catalytic proton bridge. Further experimental and theoretical works are required to elucidate this unexpected behavior.

9.3 Caging Dynamics

Solvents affect reaction dynamics of solute species by enhancing bond formation via caging reactive species on the reaction time scale and by removing the excitation energy of reactants through collisions and evaporative cooling. Cluster studies have taken an important role in investigations of solvation effects in reaction dynamics [1,112–114], due to the fact that clusters can provide a well controlled and characterized solvation environment on a molecular scale. Among the various systems studied, dihalogen molecules [115–120] and ions [121–131] have attracted special attention, a few examples are discussed here.

9.3.1 Caging Dynamics in Neutral Clusters

In an early experiment on bimolecular $I_2 \bullet Ar$ cluster [115], the fluorescence signal of I_2 following excitation above its State dissociation limit was observed and attributed to recombination in a single-atom solvent cage. Later, time-resolved caging dynamics of the I_2 molecule in large argon clusters [116,117] as well as in a high-pressure rare-gas cell [119] was investigated by Zewail and

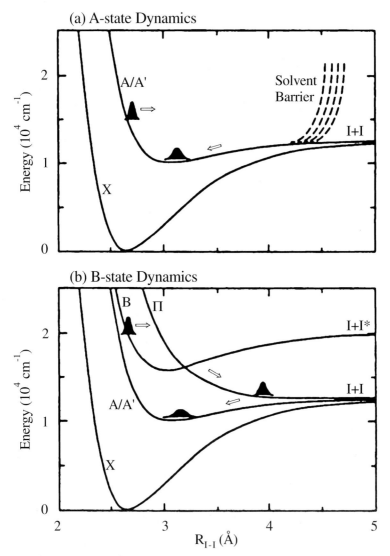

Fig. 9.4. Potential energy surfaces of the I_2 that are relevant to the dissociation and caging dynamics. (adapted from [120])

coworkers with femtosecond time-resolved spectroscopy. Molecular dynamics simulations [120] were also performed to aid in the interpretation of the data.

When the isolated iodine molecule is excited to the dissociative region of the $\tilde{A}$ state, direct dissociation occurs within 300 fs. When I_2 is solvated in Ar clusters, the photodissociated iodine fragments were observed to relax their excitation energies through collisions with the solvent cage, and subsequently recombine coherently onto the lower lying $\tilde{A}/\tilde{A}'$ or states, followed by vibrational relaxation. The recovery occurs in picosecond time scale. This coherent recovery is absent in small or warm clusters. Figure 9.4 shows a schematic of the potential energy surfaces of I_2 that are relevant to the dissociation and caging dynamics.

The dissociation and recombination dynamics of I_2 on the $\tilde{B}$ state is somewhat different. Photoexcitation of isolated iodine molecules to below the dissociation continuum of the B state will not result in dissociation. Predissociation occurs through solvent-induced curve crossing to the lower lying $\tilde{A}/\tilde{A}'$ state or the ground $\tilde{X}$ state, followed by subsequent vibrational relaxation. The predissociation rate of the $\tilde{B}$ state is solvent dependent, and occurs on a much longer timescale when compared to the direct $\tilde{A}$ state dissociation. During the predissociation process, energy flows from the photoexcited iodine molecule to the thermal vibrational modes of the cluster, loosening up the solvent cage. Therefore, the iodine fragments are able to break through the cage and separate far from each other upon dissociation. Subsequent recombination occurs through diffusion of the separated iodine fragments through the cluster, and no coherence was observed in the recombination process.

Experimental results show that the coherent caging dynamics is dependent on the dissociation time scale, the energy exchange between the iodine fragments and the cluster cage, and also the collective solvent properties such as geometry and temperature. Coherent dissociation followed by coherent recombination has also been observed in solid rare gas matrix experiments [132–134], whereas in high pressure gas and liquid solutions at room temperature, due to the lack of collective binding forces, coherent recombination was not observed.

Photoinduced dissociation and subsequent caging dynamics was also studied in methyl iodide clusters [135,136]. Solvent induced energy shift of the electronic states of methyl iodide dimers [137] and clusters [138,139] have been extensively studied. Evidence from spectroscopy studies of changes in the line shapes suggest that the ground state of methyl iodide is lowered in energy by about $500\,\mathrm{cm}^{-1}$ upon dimerization and up to $1000\,\mathrm{cm}^{-1}$ in larger clusters. No spectral shifts were observed for the Rydberg $\tilde{B}$ and $\tilde{C}$ states, indicating that these excited states are stabilized in the same way as the ground state. However, the repulsive $\tilde{A}$ state is only stabilized by $\sim 10\,\mathrm{cm}^{-1}$ upon clustering. The difference was attributed to the different strengths of the dipole–dipole interactions of the Rydberg states and the valence states with the solvent shell [138]. Since the Rydberg states and the repulsive $\tilde{A}$ state do not undergo the same amount of energy shift, the coupling between the $\tilde{A}$ state and the Rydberg states is altered upon solvation. Consequently, the lifetimes of the Rydberg states will change upon solvation, as was observed experimentally on the 10s Rydberg states in our lab [135]. When methyl iodide clusters are excited to the 10s Rydberg state, they dissociate within 200 fs via two competing processes: curve crossing to the repulsive $\tilde{A}$ state and internal conversion to the dissociation continuum of the 6s Rydberg state. Due to the caging effect, some of the dissociating CH_3 radicals and I atoms dissipate their excitation energies through collisions with the cluster cage and are able to recombine onto the 6s Rydberg state, or the lower lying ground $\tilde{X}$ state. Recombination onto the 6s Rydberg state occurs in ~ 3 ps [135].

Zewail and coworkers performed time-resolved experiments focused on the methyl iodide dimer [136]. In their experiment, a pump pulse at 277 nm was used to excite the methyl iodide to its dissociative $\tilde{A}$ state, the reaction dynamics was monitored by a probe pulse at 304 nm utilizing REMPI in a kinetic energy TOF spectrometer. Velocity gating provided the ability for the separation of fast and slow kinetic energy components of the I^+ ions formed. Pump-probe transients of the slow I^+ (ion) showed an interesting delayed rise behavior, with a time delay of 1.4 ps and a rise time of 1.7 ps. Reactions between the photofragmented I atom and the loosely bond methyl iodide were suggested to occur. The delay of 1.4 ps was interpreted as the formation time of the transition state complex $CH_3I_2^{\pm}$, whereas the rise time of 1.7 ps was a measure of the lifetime of this transition state complex, $CH_3I_2^{\ddagger}$.

9.3.2 Caging Dynamics in Anionic Clusters

Unlike studies involving neutral clusters, in which size selection is still an experimental challenge, size selection of clusters with ionic chromophores can be easily realized by standard mass spectrometric techniques. The first experiment on photoinduced dissociation and recombination dynamics of size-selected clusters was performed by Lineberger and coworkers on $Br_2^-(CO_2)_n$ clusters [121,122]. Later Lineberger and coworkers, as well as other researchers, expanded the scope of these experiments to include other dihalogen systems such as $I_2^-(CO_2)_n$ [121,127,128,131], $I_2^- Ar_n$ [124,125,127–130] and $I_2^-(OCS)_n$ [126] clusters.

Photoexcitation of I_2^- to dissociative states, e.g. the repulsive $A(^2\Pi_{g,3/2})$ or $A'(^2\Pi_{g,1/2})$ states, leads to the formation of neutral I atom and I^- ion, as shown in Fig. 9.5. When I_2^- is solvated in a cluster, photoexcitation to the dissociative state results in product ions with two different ion cores: uncaged I^- and caged I_2^- ions. In the former, the other I atom has escaped the cluster ion, whereas in the latter, the dissociated I_2^- has recombined due to the solvent caging effect. The quantum yield of the recombination (the branching ratio of the caged I_2^- products, often termed caging fraction) and the rate of recombination and subsequent relaxation are found to be strongly dependent on cluster size and structure.

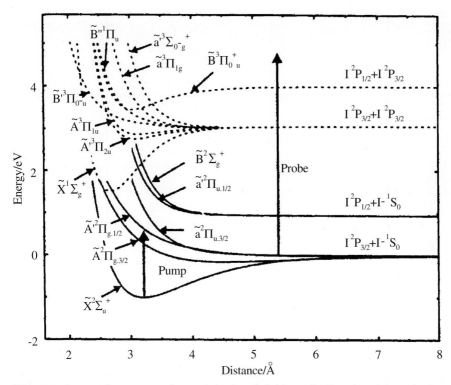

Fig. 9.5. Potential energy surfaces of the I_2^-. *Solid lines* for I_2^-; *dotted lines* for I_2. (adapted from [127])

Early photofragmentation experiments at 720 nm on $I_2^-(CO_2)_n$ clusters [121,122] showed that the caging fraction for $n \leq 5$ is zero, and reaches unity for $n \geq 16$. The onset of unity caging fraction at $n = 16$ and the magic number at $n = 16$ in the cluster distribution indicate that the first solvation shell is complete at $n = 16$. For intermediate sizes, the caging fraction increases with cluster size, as shown in Fig. 9.6a. Pump-probe absorption recovery spectroscopy was applied to study the time-dependent dynamics [124,125]. In these experiments, a picosecond pump pulse at 720 nm was used to excite mass-selected $I_2^-(CO_2)_n$ clusters to the dissociative $\tilde{A}'$ state, the absorption of a second identical probe pulse was measured as a function of the pump-probe delay time. The absorption of the probe pulse is only possible when the dissociated I_2^- has recombined/relaxed onto the $\tilde{X}$ ground state. The overall recombination and relaxation time was found to be ~ 20 ps for $n = 9$ clusters, but dropped to ~ 10 ps at $n = 16$. An interesting peak at ~ 2 ps was observed for clusters $n \geq 14$, this was attributed to the coherent I••I$^-$ motion in a rigid solvent cage on an excited region of the I_2^- potential surface. The caging dynamics of the $I_2^-(CO_2)_n$ was also studied with femtosecond 790 nm laser pulses, [124,125] which inject 150 meV less energy into the clusters. The

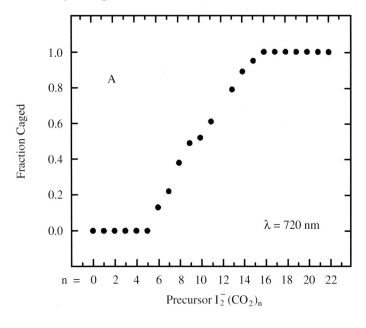

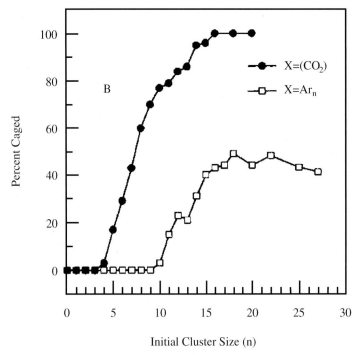

Fig. 9.6. (a) Caging fraction of the $I_2^-(CO_2)_n$ as a function of cluster size. Adapted from [122]. (b) A comparison of the caging fractions of the $I_2^-(CO_2)_n$ and $I_2^-Ar_n$ clusters as a function of cluster size. (adapted from [124])

caging fraction shows the same general dependence on cluster size as that of the 720 nm excitation, but a much faster recombination/relaxation rate was observed with lower energy 790 nm excitation. The coherent I••I$^-$ motion was also observed in the 790 nm excitation.

Caging dynamics in Ar clusters were found to be much less efficient [124,125]. At 790 nm excitation, caging was first observed at $n = 4$ for CO_2 clusters, and reaching 100% at $n = 16$; whereas in Ar clusters, caging starts at $n = 10$, and only reaches $\sim 50\%$ upon the completion of the first solvation shell at $n = 20$, as shown in Fig. 9.6b. Time-resolved absorption recovery measurements at 790 nm yield recovery times of 1.3 ps for $I_2^-(CO_2)_{16}$, and 127 ps for $I_2^-Ar_{20}$. The dramatic decrease of caging efficiency in Ar clusters is possibly due to the weaker interactions between solute-solvent and solvent-solvent. Ar has a similar mass to CO_2, so they have similar solute-solvent kinematics. Ar has no permanent dipole moment or quadrupole moment. CO_2 also has no permanent dipole moment, but it has a large electric quadrupole moment that leads to a strong interaction between I_2^-–$CO_2(\sim 200\,\text{meV})$, about four times stronger than I_2^-–Ar(53 meV). The CO_2–CO_2 interaction is also much stronger than Ar–Ar, resulting in a more rigid cage in the CO_2 case.

Introduction of femtosecond photoelectron spectroscopy (FPES) to the investigation of reaction dynamics in these ionic clusters by Neumark and coworkers [127,128] has provided additional insights into caging dynamics. In their experiments on $I_2^-(CO_2)_n$ and $I_2^-Ar_n$ clusters, a femtosecond pulse at 780 nm excites the cluster to the dissociative $\tilde{A}'$ state, and a femtosecond probe pulse at 260 nm detaches the electron from the clusters, yielding the photoelectron spectra at various time delays. FPES allows the taking of "snapshots" of the clusters evolving on different electronic and vibrational states, and also the determination of the solvent size at any delay time. Experimental results show that excitation onto the $\tilde{A}'$ state gives rise to a so-called anomalous charge-switching state in which the excited electron is localized on the less solvated I atom. For all size $I_2^-Ar_n$ clusters, dissociation is complete by 300 fs. For larger clusters ($n \geq 12$), recombination occurs after 1 ps. For $n = 12$ and 16, relaxation on the $\tilde{X}$ state was almost unobservable, whereas for $n = 20$, extensive relaxation was observed, accompanied by evaporation of all solvents by 3 ns. Relaxation on the $\tilde{A}$ state was found to be much faster, for example, at $n = 20$, it is completed by 35 ps.

In $I_2^-(CO_2)_n$ clusters, the recombination and relaxation is faster than in Ar clusters. Recombination on the $\tilde{X}$ state occurs in 10 ps for $n = 6$, and decreases to 0.5 ps for $n = 14-16$. The relaxation in the $n = 16$ cluster within the $\tilde{X}$ state is complete by 25 ps.

Investigations of photodissociation and recombination/relaxation dynamics in both neutral and ionic clusters show that solvent plays an important role in influencing the reaction dynamics of the chromophore. Even for the case of a weak solvent like Ar, the collective effect of many solvent atoms can strongly affect the reaction dynamics. In stronger bonded solvent like CO_2, the solvent effect is much more pronounced. Solvent effects are many fold: solvent molecules can cage in the dissociating fragments and remove the excitation energy of the chromophore through collisions and evaporative cooling; solvent molecules can also perturb the electronic states of the chromophore, thus changing the coupling between different electronic states.

9.4 Coulomb Explosion Process in Clusters

Ionization involving multiple electron loss was first observed in cluster systems by electron impact ionization techniques [140–145]. In most of these experiments, clusters containing one, two, or three charge centers were produced, but were only observable in systems that remained metastable. Metastability is present in systems of sufficient size that the charge centers were separated by a large enough distance that Coulomb repulsion was reduced to a magnitude that did not exceed the cohesive binding energy of the cluster. Fission occurs when the cluster is below a critical size. Much higher levels of multiple electron loss can be achieved by photoionization in an intense laser field [146–161]. With recent advances in ultrafast laser technology, table-top high peak-power laser systems capable of producing focused pulses with a power density in the range of 10^{14}–10^{19} W/cm^2 are routinely available. This advance prompted extensive experimental [146–161] as well as theoretical [162–173] interest in investigations of these multiple ionization processes.

9.4.1 Role of Clusters in the Coulomb Explosion Process

When a small molecule is irradiated by an intense ultrafast laser pulse, all atoms in the molecule can be ionized. Thereafter, the molecule rapidly explodes due to the Coulomb repulsion of like charges, releasing atomic ions with a range of kinetic energies. The kinetic energy release of the atomic ions from the Coulomb explosion (CE) of small molecules is relatively small, being in the order of several to tens of electron volts [146–148]. In clusters, multiple ionization in an intense laser field can lead to ionization of all cluster atoms, consequently resulting in a much more violent CE process, releasing atomic ions with much higher kinetic energies and charge states. It is expected that the highest observed charge-state and kinetic energy of the Coulomb exploded atomic ions increase with the size of the system. Atomic ions with kinetic energies of several hundreds to thousands of eV and high charge- states of up to +20 were first observed in our lab [149–152], when van der Waals and hydrogen bonded molecular clusters ($n \sim 10^2$) were irradiated in an intense laser field with a power density of only $\sim 10^{15}$ W/cm^2. When larger clusters, $n = 10^3 \sim 10^5$ molecules, were exposed to a laser field with an intensity of around 10^{16} W/cm^2, atomic ions with even higher kinetic energies, in the range of keV, even up to MeV, were observed [153,154].

Various experimental observations show that with moderate laser intensity, high values of kinetic energy release and charge-state can only be observed when clusters are involved [149–156]. In order to prove the role clusters play in the CE process, we employed a covariance analysis technique [174], in which various cluster distributions in a pulsed molecular beam were probed, and the high values of kinetic energy and charge-state of the atomic ions were correlated with cluster size.

In most photoionization mass spectroscopic studies, ion intensities are averaged over many laser shots before analysis and interpretation. For a single shot measurement, deviations in the ion intensity are unavoidable due to fluctuations in the laser and molecular beam source. Covariance analysis takes advantage of these fluctuations and compares the changes in one measurement

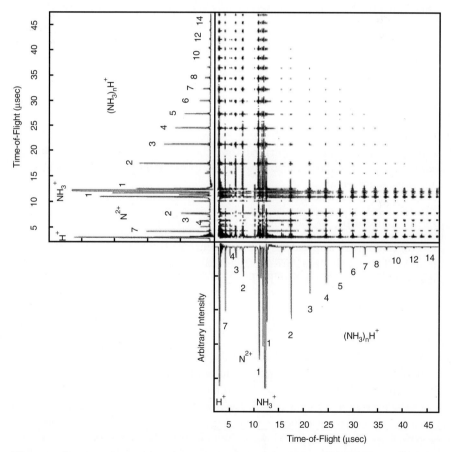

Fig. 9.7. Anticovariance map of ammonia clusters under laser irradiation of 624 nm, 120 fs, and $\sim 10^{15}$ W/cm^2. The abscissa and ordinate display a TOF spectrum of ammonia clusters averaged over 10 000 single-shot spectra. (adapted from [174])

with respect to another via a shot-by-shot analysis, thus providing a measure of the connectivity between two different reaction channels [174].

Covariance is defined as the relative deviation of two different species in two measurements, just as variance is the deviation of the same species in two measurements. In a simplified view of mass spectroscopic studies of chemical reactions, two mass peaks are correlated if both either increase or decrease in intensity in different measurements. On the other hand, they are anti-correlated if one increases while the other decreases. And if one peak remains unchanged while the other increases or decreases, they are un-correlated. For a full description of the covariance analysis, see [174].

Our covariance analysis of the CE process of ammonia clusters induced by an intense laser field shows that there is a strong anti-covariance between large clusters and the Coulomb exploded nitrogen ions, as shown in Fig. 9.7. This indicates that the production of Coulomb exploded nitrogen ions is at the expense of the large ammonia clusters. This result provides strong evidence that the highly charged Coulomb exploded species originate from clusters.

This conclusion was also confirmed by another series of experiments [155,156], in which the delay between the laser pulse and the molecular beam was scanned in order to favor the monomer, small clusters, or large clusters. Charge state distribution, kinetic energy release, and integrated signal intensity, as a function of the delay between the laser and molecular beam were examined. Results confirmed that the high charged species with high values of kinetic energy release were only present when clusters dominated the molecular beam.

9.4.2 Modeling of Coulomb Explosion Process

Several theoretical approaches have been introduced to model the CE process in molecules and clusters [157,158,162–173]. Among them are the charge resonance enhanced ionization (CREI), coherent electron motion model (CEMM [157–160]) and ionization ignition model (IIM [169,170]). In all these models, barrier suppression plays an important role in the first stage of cluster ionization. For a laser beam with an intensity of $\sim 10^{15}$ W/cm^2, the field strength is on the order of 10 V/Å, which is strong enough to field ionize at least one of the valence electrons of all the nuclei in the cluster. In the IIM picture [169,170], at the very beginning of the ionization events, the nuclei in the cluster are considered frozen while the ionized electrons are quickly removed from the cluster by the laser field. Even in small clusters, the unshielded ion cores in the cluster produce a very large (> 102 V/Å) and inhomogeneous electric field. This large field lowers the ionization barrier and facilitates subsequent ionization processes, which further lowers the ionization barrier. The field created by the initial ionization event appears to "*ignite*" the cluster, which undergoes further ionization processes. The IIM model predicts that multiple ionization is strongly dependent on the cluster density, but is not very sensitive to either the atomic weights of cluster constituents, or the pulse duration of the laser beam, as long as the same intensity can be obtained.

In the CEMM model [157–160], interaction between the laser field and clusters can enter a regime of strong electromagnetic coupling which arises from the *coherent motion* of the field ionized electrons (Z) induced by the external laser field. These coherently coupled electrons behave like a quasi-particle with a charge Ze and a mass Zm_e. Subsequent ionization can be envisioned as the electron impact ionization by these coherently energized electrons. With the inclusion of this coherent electron motion, Rhodes and coworkers were able to explain their experimentally observed L- and M-shell X-ray emission from Xe and Kr clusters ($n \sim 100$) exposed to laser powers in the range of 10^{16}–10^{19} W/cm^2. According to the CEMM model, the ionization rate of the cluster increases with cluster size, but decreases with internuclear distances. This is in contrast to the CREI model [162–170], which predicts that the rate of ionization is a highly irregular function of the internuclear distance.

In the CREI model for diatomic molecules [162–168], an inner potential barrier existing between the two nuclei rises with the internuclear distances. The electron motion between the two nuclei slows down due to the barrier. At a certain critical distance, when the characteristic frequency of the electron motion become in resonance with the laser, the electron energy is enhanced, leading to an enhanced ionization rate. A dynamic CREI model was later proposed by Jortner and coworkers [171,172], in which the electron energy was shown to increase roughly at the same rate as the level of the barrier height. The dynamic model predicts a higher ionization efficiency than the static CREI model.

In our experiments [149,152,155,156] on small clusters, multi- charged atomic ions resulting from the CE process were observed only when a certain minimum cluster size was reached, and the charge distribution does not vary significantly with further variation in the cluster distribution. With a laser power of $\sim 10^{15}$ W/cm^2, pure xenon, ammonia, and acetone clusters readily undergo the CE process, whereas krypton and argon clusters do not, unless a trace amount of HI or other low IP molecule is doped in the clusters. The fact that under the laser power employed in our experiments, neat Kr (IP = 14.00 eV) and Ar (IP = 15.76 eV) clusters do not exhibit multiple ionization, whereas neat Xe (IP = 12.13 eV), HI (IP = 10.39 eV), NH$_3$ (IP = 10.18 eV), and (CH$_3$)$_2$CO (IP = 9.71 eV) clusters do, suggests that multiple ionization of clusters is less sensitive to atomic number of the cluster constituents and depends more upon the threshold for single ionization. These results agree with the IIM model predictions.

In a one-color pump-probe experiment of acetone clusters (the pump pulse at 624 nm has a power density of $\sim 3 \times 10^{14}$ W/cm^2 and the probe pulse is (10% weaker), transients of the oxygen ions O^{n+} ($1 \leq n \leq 5$) display unusual asymmetric beating patterns with phase shifts (Fig. 9.8). The unequal inten-

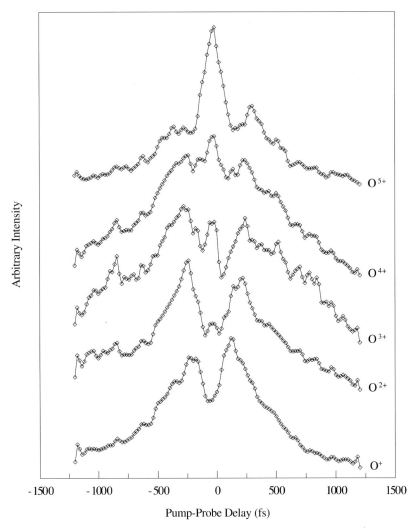

Fig. 9.8. Pump-probe transients of multicharged oxygen ions, O^{n+} ($1 \leq n \leq 5$), formed through the CE of the acetone clusters. (adapted from [152])

sities of the maxima at positive and negative delay times can be attributed to the unequal pump and probe intensities, but significantly, the maxima in O^+ and O^{5+} occur at positive delay times, whereas the maxima in O^{2+}, O^{3+}, and O^{4+} occur at negative delay times. The C^{n+} ($1 \leq n \leq 4$) ions observed were found to behave in a similar fashion. Obviously, the observed beating pattern could not arise from the phasing of the pump and probe optical fields since the period of the laser is about 2 fs for 624 nm laser pulses. Bear in mind that the power of either pump or probe beam is sufficient to multiply ionize the acetone clusters. Following the pump pulse, acetone cluster undergoes multiple electron loss, and the probe beam arrives at the cluster at some later time. Depending on the time delay of the probe beam, the interatomic spacing has increased to a particular distance as a result of the nuclear motion arising from the CE process. As the interatomic spacing is varied, the wavefunction and electron localization is changed. According to the CREI model, this would result in different ionization rates for the various charge states, thus yielding the beating pattern seen in Fig. 9.8.

Our studies of van der Waals and hydrogen bonded clusters irradiated with intense laser fields [149,152,155,156] demonstrate that the high charge states obtained are well described by the IIM and CREI model. The lack of dependence on atomic weight and the strong dependence on ionization potential and internuclear distances, support this conclusion. Present results cannot totally eliminate the possibility of coherent electron motion, but under experimental conditions of intermediate laser power and small cluster size, it is not a major contribution to the multiple ionization and subsequent CE process.

9.4.3 Coulomb Explosion Imaging

Molecules and clusters have a propensity to undergo CE processes when exposed to intense light fields available from ultrafast laser pulses. In our lab, we have invented a new technique – Coulomb explosion imaging, which takes advantage of this phenomenon to gain information on the time evolution of the reaction intermediates [15,17]. In the Coulomb explosion imaging technique, a first femtosecond laser pulse is employed to launch the system to an excited state of interest. A second intense femtosecond laser pulse is used to initiate the CE process in the reaction intermediates, and the Coulomb exploded fragments are detected at defined delay times. Calculations [175] have placed the approximate separation times of the multiple charged species in the neighborhood of 25 fs, so that little change occurs after the initial CE event. Therefore, information regarding the reaction intermediates can be extrapolated from the time-resolved detection of fragments. This technique has been successfully applied in the study of the tautomerization reaction in the 7-azaindole dimer [15,17], as shown schematically in Fig. 9.9. In this experiment, a low power pump pulse is used to excite the 7-azaindole dimer to its S_1 state, where it undergoes double proton transfer; a second intense probe pulse multiple ionizes the evolving dimer at a prefixed delay time, leading to its fragmentation. The detection of the resulting fragment ions as a function of delay time between the pump and probe laser pulses depends on the progression of the proton transfer process. By monitoring the time evolution of the fragment ion with $m/z = 119$ (produced by CE of the intermediate state after the first proton transfer step, see Fig. 9.9) and other mass peaks, a two step proton transfer mechanism was confirmed, supporting the previously obtained results of Zewail and coworkers [106].

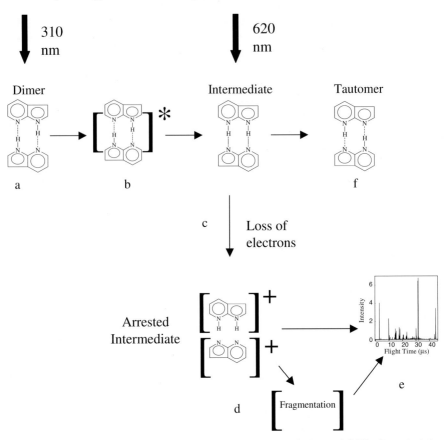

Fig. 9.9. Mechanism for Coulomb explosion imaging technique. (**a**) The 7-azaindole dimer is excited by a 310 nm femtosecond pulse. (**b**) The excited half of the 7-azaindole initiates the tautomerization process. (**c**) Coulomb exploding 620 nm pulse separates the dimer in ∼ 25 fs to freeze the reaction intermediate. (**d**) The intermediate fragment (119 amu) is ionized in the CE process while the 117 amu half is unstable and fragments into several low-mass charged species. (**e**) Cations are detected in a time-of-flight mass spectrometer. (**f**) Full tautomerization proceeds if the 7-azaindole dimer is not interrupted by Coulomb exploding femtosecond radiation. (adapted from [17])

In addition to the Coulomb explosion imaging technique, which shows potential applications in monitoring structural changes in real time, irradiation of clusters with intense laser field offers other potential applications as well. For example, the generation of ultrafast X-ray pulses, and table-top fusion experiments and the resulting production of neutron beam. Rhodes and coworkers have measured very efficient X-ray emission in rare gas clusters excited with intense laser field. They predicted that the emission time for kilovolt hard X-rays could be sub-picosecond, and control of the pulse width and wavelength of the X-ray may be possible through molecular design [176]. These findings may lead to the ultimate development of ultrafast X-ray lasers, which could provide real-time imaging with molecular resolution. Ditmire and coworkers [153,154] observed the $D+D \to He_3 + n$ fusion reaction, when high-energy deuterium ions ($>$ keV) were created by the interaction of an intense laser field ($\sim 2 \times 10^{16}$ W/cm^2) with deuterium clusters. An efficiency of about 10^5 fusion neutrons per joule of incident laser energy was obtained, which approaches the efficiency in large scale laser-driven fusion experiments. This result shows the possibility of developing table-top fusion experiments using small scale laser systems, which could lead to the development of table-top neutron source and may find a wide range of applications in various fields.

9.5 Electronic Excitation, Relaxation and Ionization of Met-Cars

Studies of matter comprised of clusters is particularly revealing in terms of obtaining a more complete understanding of factors governing the properties of systems of finite size, since they provide the opportunity, for example, to investigate when discrete molecular properties begin to coalesce and display the collective behavior characteristic of extended solids (e.g., band structure). Small systems often display optical absorption features attributable to quantum confinement and sometimes show behavior that can span semiconductor to metallic-like characteristics. A related subject of considerable current interest is determining the molecular details of the phenomenon of delayed ionization in small systems, which has thermionic emission as its bulk phase analogue, and also ascertaining how the electronic states of mixed transition metal containing clusters couple in terms of their ionization characteristics. Of significance are questions addressing whether there is a behavioral difference attributable to the time domain of the ionization source, i.e., nanosecond versus femtosecond electromagnetic radiation. Answering these questions for systems comprised of metal compound clusters will have a significant impact on understanding the properties of new materials, and the prospects of using clusters to tailor design new nanoscale materials with desired properties.

Several years ago during the course of undertaking detailed studies of dehydrogenation reactions of hydrocarbons induced by titanium ions, atoms, and clusters, we discovered the formation of an unusually abundant and stable cationic species having a molecular weight of 528 amu, which we thereafter established contained eight titanium atoms and twelve carbons [18–21]. Subsequent work revealed the existence of the neutral molecular cluster, its anion analogue, the stability of other transition metal-carbon complexes of identical stoichiometry, and thereafter a general class of caged molecular clusters comprised of early transition metals bound to carbon atoms in the same stoichiometric ratio [18–64]. These have been termed metallocarbohedrenes, or Met-Cars for short. Discussing some of their unique electronic properties is the subject of this section.

9.5.1 Met-Cars: A Unique Molecular Cluster System

The first Met-Car molecular cluster ion was generated through reactions of titanium with vapors of a variety of small hydrocarbon molecules, utilizing a laser-based plasma reactor. The mass distributions first obtained with methane and acetylene showed a peak corresponding to 528 amu to be completely dominant, while there are not any other prominent peaks observable in the mass range below 1200 amu. Reactions with ethylene, benzene, and propylene, also generated similar cluster distributions, as did ones conducted with rods of mixed Ti and graphitic powders [177].

Following these observations, we first undertook a series of studies with hydrocarbons of varying isotopic composition in order to definitively establish the identity of this unusually stable species. Isotope labeling experiments with deuterated organic compounds did not display any shift in mass, showing that the clusters corresponding to the peak at 528 amu did not contain any hydrogen atoms. ^{13}C labeling experiments established that the cluster accommodated exactly twelve carbon atoms. Subsequent high-resolution isotope distribution pattern analyses supported the presence of eight Ti atoms, leading to the assignment of $Ti_8C_{12}^+$. Later studies also revealed that it was possible to assemble Met-Cars of binary composition among the early transition metals such as Zr and Hf, as well as with some other atoms which do not form the pure Met-Car species, i.e., Y, Ta, W, and Si [24–26].

Early findings based on titration experiments showed a smooth uptake of eight bonded ligands, implying that each of the titanium atoms were bonded to three carbon atoms through Ti–C bonds, and that each of the carbons were also bound to its adjacent carbon through a double bond [40–42,46,47,178–181]. This led to the suggestion that this species (as well as its corresponding neutral molecule discovered later) might have a pentagonal dodecahedron structure. Since then a variety of structures in addition to the pentagonal dodecahedron have been considered theoretically [43–45,48–50,54,183], and more recent experiments and theory tend to support a T_d structure [64,185] rather than the one originally proposed which has T_h symmetry [18].

9.5.2 Delayed Ionization

The phenomenon of delayed ionization in clusters has been reported for several systems including the Met-Cars [56,57,61], pure transition metals [186–188], metal oxides [189–191], metal carbides [191], and the fullerenes [192–194]. Recent developments in the case of fullerenes are discussed in the chapter by Campbell and Levine. Currently there is considerable interest in the origin of this phenomenon, prompting extensive theoretical and experimental investigations for various systems [61,62,195]. In order for a cluster to display delayed ionization in a manner analogous to bulk phase thermionic emission, the ionization potential of the cluster must be less than its dissociation energy; and all of phase space must be accessed by the system. The first requirement ensures that ionization, as an energy dissipation mechanism, will be more favorable than dissociation. Theoretical calculations and experimental results show that Met-Car clusters meet this requirement. The second requirement enables the system to temporarily store energy in excess of the cluster's ionization potential through statistically sampling a large number of accessible vibrational and electronic states.

Studies conducted in our laboratory employing nanosecond lasers revealed that some degree of ionization occurred on a time scale orders of magnitude longer than that which is characteristic of normal photoionization that obeys the photoelectric effect. In order for the delayed ionization to be observed for Met-Cars, the clusters must have some way to accommodate the energy necessary for the ionization to occur, while at the same time not undergo dissociation into smaller cluster fragments. Metallocarbohedrenes are thus ideal systems to exhibit this behavior, because when comparing the experimentally measured ionization potential (IP) and the theoretically predicted value for the dissociation energy (E_{diss}), a favorable relationship (IP/$E_{\mathrm{diss}} < 1$) exists for this family of cluster molecules. This favorable relationship and the large density of electronic states for these transition metals-carbon species, may allow for the clusters to "store" the energy gained during the excitation and delay ionize on a long time scale characteristic of the experiment. In the case of titanium-carbon clusters, delayed ionization was observed only for Met-Cars, but not for the remainder of the metal-carbon cluster distribution [61]. At high laser fluences($> 50\,\mathrm{mJ/cm^2}$), a second delayed channel which corresponds to an atomic ion emission was observed [56,57]. Neither of these delayed ion channels, however, exhibit a dependence on the laser excitation wavelengths of 532 and 355 nm, providing evidence that excitation to a specific (e.g. triplet) state is not responsible for the observed delayed ionization [61]. Each channel did display a strong dependence on the fluence of the excitation laser. Similar findings have been consistently seen in all of the Met-Car systems studied, including Ti_8C_{12}, V_8C_{12}, Zr_8C_{12}, as well as the binary metal containing Met-Cars $Ti_xZr_yC_{12}$ and $Ti_xNb_yC_{12}$ (where $x+y=8$). Further, a delayed atomic ion emission channel has been observed only for the Ti^+ component in the binary metal Met-Car systems [57].

Even in those metal-carbon systems that display delayed ionization for cluster compositions other than those of the Met-Car stoichiometry, in all of the pure metal systems investigated (Ti, Zr, Nb), the cluster corresponding to the Met-Car stoichiometry was found to be the dominant species to exhibit delayed ionization at all wavelengths investigated [61]. It should be noted that for single metal Met-Cars, the rate of delayed ionization is dependent on the type of metal incorporated into the cluster, while in the mixed metal

Met-Cars, a non-monotonic dependence was found with respect to the degree of metal substitution [57].

The general process of delayed ionization is most readily treated through use of the Richardson-Dushman equation which is an expression derived for the macroscopic bulk phenomenon often observed in metallic systems. The equation has been reformulated by Klots to take into account the finite size of the system under consideration and the collection of excited species having a distribution of internal energies [195]. The data for the mixed metal Met-Cars show a non-monotonic behavior with respect to metal substitution, and are found to correlate well with the ionization potential measured in our laboratory. The observed temporal dependence for both the pure and mixed metal Met-Cars is in good agreement with the Klots model [56,57,61].

Despite the satisfactory agreement between the Klots model and the observations, our calculations [61] reveal the presence of various electronic bands in these transition metal-carbon cluster systems. In such a case, a deviation in the macroscopic bulk-like behavior would be expected. These considerations, together with our interest in further unraveling the molecular aspects of the phenomenon, prompted us to undertake a detailed study of the ionization dynamics of Met-Cars using femtosecond pump-probe techniques.

9.5.3 Ultrafast Spectroscopy

Observation of significant delayed ionization in the Met-Cars clusters and the fact that these clusters have unexpectedly low ionization potentials has stimulated considerable interest in elucidating dynamical aspects of their electronic properties. Although Met-Cars have attracted considerable interest from the theoretical community, to date many details concerning their electronic structure are not well known due to the complexity associated with calculations on these systems which have numerous electrons.

As a contribution toward more fully elucidating the electronic properties of Met-Cars, a series of experiments were performed in our laboratory on the excitation and relaxation dynamics of various vanadium-carbon clusters [58,60]. Particular emphasis was placed on determining the influence of laser fluence as well as laser wavelength on the ionization behavior of these clusters. In addition to determining the time-resolved dynamics of electronic relaxation in various vanadium-carbon clusters, data revealing nuclear vibrational motion also has been acquired.

In one series of measurements, the femtosecond dynamics of vanadium-carbon clusters were investigated employing 400 nm pump and 620 nm probe. The pump-probe responses of the V_8C_{12} cluster was determined to be 225 fs (FWHM), which is much longer than the cross-correlation of the pump-probe beam (≤ 86 fs), indicating that a state (or band of states) with an appreciable lifetime is being accessed. Fitted transients for a series of vanadium-carbon clusters, including V_8C_{12}, are presented in Fig. 9.10. The Met-Car response is noticeably longer than the autocorrelation width and, since these clusters

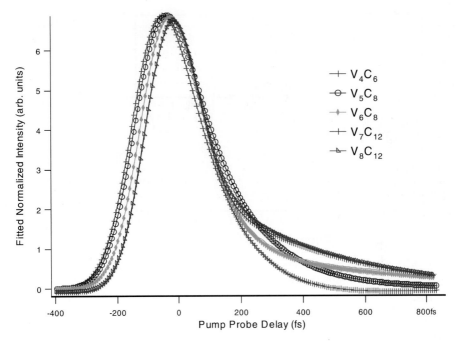

Fig. 9.10. Summary of the fitted pump-probe transients of the vanadium-carbon clusters: V_4C_6, V_5C_8, V_6C_8, V_7C_{12}, and V_8C_{12}. Pump: 400 nm, 50 fs, 24 mJ; probe: 620 nm, 50 fs, 20 mJ. (adapted from [58])

are strongly bound [58], the pump-probe response is likely too short to be attributable to a fragmentation process, either in the Met-Car cluster, or in larger clusters that might be thought as possible contributors to the formation of a Met-Car. The pump-probe response is thus attributed to the temporal dependence of the electronic relaxation behavior of this system. Another significant observation is the fact that the pump-probe transients for all of the observed vanadium-carbon clusters with more than four metal atoms show a similar pump-probe response. They display exponential decays on the order of a few hundred femtoseconds. This observation suggests that there is a common chromophore [58] present in clusters, likely to be the V_4C_x series. It is also important to note that the excited state lifetimes become shorter as the cluster size increases. This result suggests that the larger nanoscale complexes have increasingly more free-electron character and therefore the rate of energy relaxation is higher in the larger clusters. The long-time tail in the Met-Car spectrum suggests relaxation into a band or band of states.

Similar relaxation behavior is observed at other wavelengths as well. Figure 9.11 shows a typical pump-probe transient of the vanadium Met-Car at moderate laser fluence with pump pulse at 1260 nm and probe pulse at 800 nm. The experimental findings begin to reveal the large density of excited electronic states involved, and the concomitant effect on the relax-

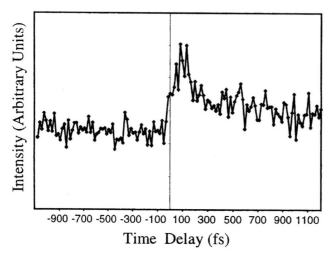

Fig. 9.11. Pump-probe transient of the V_8C_{12} Met-Car observed at 1260 nm pump and 800 nm probe. (adapted from [59])

ation dynamics. When the pump pulse is tuned slightly to redder wavelength, the relaxation dynamics changed dramatically, as shown in Fig. 9.12. Clear evidence for electronic energy coupling to lattice modes has been acquired, with findings that at long delay times the latter could contribute to the thermionic emission we have observed. Importantly, these new findings show

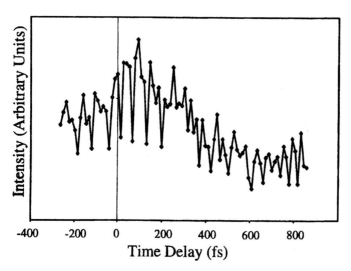

Fig. 9.12. Pump-probe transient of the V_8C_{12} Met-Car observed at 1300 nm pump and 800 nm probe. Note the oscillatory frequency: $505\,\text{cm}^{-1}$. (adapted from [60])

that the delayed ionization process, typically seen only under excitation via nanosecond pulses, might be revealed in experiments conducted with laser fluences that enable relaxation into electronic bands during a femtosecond excitation process. This work is providing a new understanding of the process generally termed "thermionic" ionization/emission, in terms of the microscopic quantum effects of the influence of the relaxation dynamics into electronic bands that govern the temporal characteristics of the phenomena.

Other recent preliminary investigations exploring the electronic excitation and relaxation properties of the vanadium Met-Car, using short unchirped pulses, have also enabled us to observe and characterize vibrational frequencies involved in the ionization dynamics of the vanadium Met-Car (see Fig. 9.12). Periodic recurrences in the ionization intensities are seen, from which a frequency of approximately $505\,\mathrm{cm}^{-1}$ is obtained [60,196]. Interestingly, this compares to a metal-carbon stretch frequency of $575\,\mathrm{cm}^{-1}$ reported for VC on vanadium surfaces [197]. It should be noted that Meijer and coworkers [198] have extracted a similar vibration for Ti_8C_{12} with a frequency of $525\,\mathrm{cm}^{-1}$.

9.6 Conclusion

Cluster research is an active topic of investigation in the field of chemical physics, motivated in part by the realization of its value in bridging an understanding of phenomena in the gas and condensed phases by providing opportunities of studying the reaction dynamics at controlled degrees of solvation.

In this chapter, we have demonstrated that clusters can serve as model systems in studies involving charge-transfer, photodissociation, recombination and subsequent relaxation processes. The discussions were focused to a large extent on work performed in our laboratory, special consideration was given to investigations utilizing femtosecond pump-probe spectroscopy. The findings discussed herein are a testimony to the wide-ranging processes that can be explored within the domain of femtosecond pump-probe spectroscopy. Of particular interests are when the clusters are exposed to intense laser field, they undergo CE giving rise to particles with high charge states and surprisingly large values of kinetic energy release. In addition to the fundamental interest in the CE phenomenon, the development of an interesting new technique, Coulomb explosion imaging, has emerged that enables the intermediates in the ultrafast reactions to be arrested and investigated as they undergo transformation from reactants to products via the intermediate states.

The various findings reviewed in this chapter clearly demonstrate the power of time-resolved spectroscopy in unraveling the dynamics of clusters, and we can expect to observe increasing activity in this field of endeavor. With the availability of tunable table-top solid-state laser systems that can routinely deliver focused powers in 10^{15}–10^{19} W/cm^2 range and have time-resolutions in femtosecond, even sub-femtosecond regime, many more exciting discoveries are yet to come.

Acknowledgements. We gratefully acknowledge financial support by U.S. Air Force Office of Scientific Research Grant F49620-01-1-0122 (AASERT01), the U.S. Department of Energy Grant DE-FG02-97ER14258, the U.S. National Science Foundation CHE-99-06341, and U.S. National Science Foundation ATM 97-11970. QZ acknowledges the postdoctoral fellowship from the NRL-NRC Research Associateship Programs.

References

1. Castleman, Jr., A.W.; Bowen, Jr., K.H., J. Phys. Chem. **100**, 12911, (1996).
2. Castleman, Jr., A.W. Advances in Mass Spectrometry, **12**, 167, Int. J. Mass Spectrom. Ion Proc. 1992, 118, 167.
3. Castleman, Jr., A.W.; Wei, S. Ann. Rev. Phys. Chem. , **45**, 685, (1994).
4. Castleman, Jr., A.W. Int. J. Quantum Chemistry **25**, 527 (1991).
5. Castleman, Jr., A.W.; Keesee, R.G. Chem. Rev. **86**, 589 (1986).
6. Castleman, Jr., A.W.; Keesee, R.G. Science, **241**, 36 (1988).
7. Clusters of Atoms and Molecules; Haberland, H., Ed., Springer-Verlag, Berlin Heidelberg, 1994.
8. Q. Zhong, A.W. Castleman, Jr., Chem. Rev., **100**, 4039 (2000).
9. S. Wei and A.W. Castleman, Jr., in Chemical Reactions in Clusters (E.R. Bernstein, Ed.) Oxford University Press, pp. 197–220 (1996).
10. Echt, O.; Dao, P.D.; Morgan, S.; Castleman, Jr., A.W. J. Chem. Phys. **82**, 4076, (1985).
11. Mamyrin, B.A.; Karataev, V.I.; Shmikk, D.V.; Zagulin, V.A. Sov. Phys. JETP **37**, 45 (1973).
12. R. Weinkauf, K. Walter, C. Weichkhardt, U. Boesl, E.W. Schlag, Z. Naturforsch. **44a**, 1219 (1989).
13. Femtochemistry – Ultrafast Dynamics of The Chemical Bond, Zewail, A. H., World Scientific (20th Century Chemistry Series), New Jersey, Singapore 1994; Femtosecond Chemistry, Manz, J.; Wöste, L. (Eds.), VCH Verlagsellschaft: Weinheim 1995; Zewail, A.H., J. Phys. Chem. **100**, 12701, (1996).
14. Dantus, M.; Rosker, M.J.; Zewail, A.H., J. Chem. Phys. 87, 2395, (1987); Zewail, A.H., Science **242**, 1645, (1988).
15. Folmer, D.E.; Poth, L.; Wisniewski, E.S.; Castleman, Jr. A.W., Chem. Phys. Lett. **287**, 1, (1998).
16. D.E. Folmer, E.S. Wisniewski, S.M. Hurley, and A.W. Castleman, Jr. Proc. Natl. Acad. Sci., **96**, 12980, (1999).
17. Folmer, D.E.; Wisniewski, E.S.; Castleman, Jr., A.W., Chem Phys. Lett. **318**, 637, (2000).
18. Guo, B.C.; Kerns, K.P.; Castleman, A.W., Jr., Science **255**, 1411, (1992).
19. Guo, B.C.; Wei, S.; Purnell, J.; Buzzza, S.A.; Castleman, A.W., Jr. Science, **256**, 515, (1992).
20. Wei, S.; Guo, B.C.; Purnell, J.; Buzza, S.A.; Castleman, A.W., Jr. J. Phys. Chem. **96**, 4166, (1992).

21. Wei, S.; Guo, B.C.; Purnell, J.; Buzza, S.A.; Castleman, A.W., Jr. Science **256**, 818, (1992).
22. Pilgrim, J.S.; Duncan, M.A., J. Am. Chem. Soc. **115**, 6958, (1993).
23. Guo, B.C.; Castleman, A.W., Jr. In Advances in Metal and Semiconductor Clusters; Duncan, M., Ed.; JAI Press: 1994; Vol. 2, pp. 137–164.
24. Cartier, S.F.; May, B.D.; Castleman, A.W., Jr. J. Chem. Phys. **100**, 5384, (1994).
25. Cartier, S.F.; May, B.D.; Castleman, A.W., Jr. J. Am. Chem. Soc. **116**, 5295, (1994).
26. Deng, H.T.; Guo, B.C.; Kerns, K.P.; Castleman, A.W., Jr. Int. J. Mass Spectrom. Ion Processes **138**, 275, (1994).
27. Guo, B.C.; Kerns, K.P.; Castleman, A.W., Jr. J. Am. Chem. Soc. **115**, 7415, (1993).
28. Wei, S.; Guo, B.C.; Purnell, J.; Buzza, S.A.; Castleman, A.W., Jr. J. Phys. Chem. **97**, 9559, (1993); **98**, 9682, (1994).
29. Cartier, S.F.; May, B.D.; Toleno, B.J.; Purnell, J.; Wei, S.; Castleman, A.W., Jr. Chem. Phys. Lett. **220**, 23, (1994).
30. Wei, S.; Guo, B.C.; Deng, H.T.; Kerns, K.; Purnell, J.; Buzza, S.A.; Castleman, A.W., Jr. J. Am. Chem. Soc. **116**, 4475, (1994).
31. Wei, S.; Castleman, A.W., Jr. Chem. Phys. Lett. **227**, 305, (1994).
32. Purnell, J.; Wei, S.; Castleman, A.W., Jr. Chem. Phys. Lett. **229**, 105, (1994).
33. Kerns, K.P.; Guo, B.C.; Deng, H.T.; Castleman, A.W., Jr. J. Am. Chem. Soc. **117**, 4026, (1995).
34. Deng, H.T.; Guo, B.C.; Kerns, K.P.; Castleman, A.W., Jr. J. Phys. Chem. **98**, 13373, (1994).
35. Kerns, K.P.; Guo, B.C.; Deng, H.T.; Castleman, A.W., Jr. J. Chem. Phys. **101**, 8529, (1994).
36. Deng, H.T.; Kerns, K.P.; Castleman, A.W., Jr. J. Am. Chem. Soc. **118**, 446, (1996).
37. Cartier, S.F.; May, B.D.; Castleman, A.W., Jr. J. Phys. Chem. **100**, 8175, (1996).
38. Lee, S.; Gotts, N.G.; von Helden, G.; Bowers, M.T. Science **267**, 999, (1995).
39. Yeh, C.S.; Afzaal, S.; Lee, S.A.; Byun, Y.G.; Freiser, B.S. J. Am. Chem. Soc. **116**, 8806, (1994).
40. Grimes, R.W.; Gale, J.D. J. Chem. Soc., Chem. Commun. 1222, (1992).
41. Reddy, B.V.; Khanna, S.N.; Jena, P. Science **258**, 1640, (1992).
42. Rantala, T.; Jelski, D.A.; Bowser, J.R.; Xia, X.; George, T.F. Z. Phys. D **26**, 5255, (1993).
43. Ceulemans, A.; Fowler, P.W. J. Chem. Soc., Faraday Trans. **88**, 2797, (1992).
44. Rohmer, M.; De Vaal, P.; Bénard, M. J. Am. Chem. Soc. **114**, 9696, (1992).
45. Lin. Z.; Hall, M.B. J. Am. Chem. Soc. **114**, 10054, (1992).
46. Methfessel, M.; van Schilfgaarde, M.; Scheffler, M. Phys. Rev. Lett. 70, 29, (1993); Phys. Rev. Lett. **71**, 209, (1993).
47. Grimes, R.W.; Gale, J.D. J. Phys. Chem. **97**, 4616, (1993).
48. Dance, I.J. Chem. Soc., Chem. Commun. 1779, (1992).
49. Chen, H.; Feyereisen, M.; Long, X.P.; Fitzgerald, G. Phys. Rev. Lett. **71**, 1732, (1993).
50. Rohmer, M.-M.; Bénard, M.; Henrite, C.; Bo, C.; Poblet, J.-M. J. Chem. Soc., Chem. Commun. 1182, (1993).

51. Dance, I., G. Aust. J. Chem. **46**, 727, (1993).
52. Rohmer, M.-M.; Bénard, M.; Bo, C.; Poblet, J.-M. J. Am. Chem. Soc. **117**, 508, (1995).
53. Khan, A.J. Phys. Chem. **99**, 4923, (1995).
54. Pauling, L. Proc. Natl. Acad. Sci. U.S.A. **89**, 8175, (1992).
55. Cartier, S.F.; Chen, Z.Y.; Walder, G.J.; Sleppy, C.R.; Castleman, A.W., Jr. Science **260**, 195, (1993).
56. May, B.D.; Cartier, S.F.; Castleman, A.W., Jr. Chem. Phys. Lett. **242**, 265, (1995).
57. Cartier, S.F.; May, B.D.; Castleman, A.W., Jr. J. Chem. Phys. **104**, 3423, (1996).
58. S.E. Kooi, B.D. Leskiw, and A.W. Castleman, Jr. Nano Letters **1**, 113, (2001).
59. E. Wisniewski, B. Leskiw, S. Hurley, T. Dermota, D. Hydutsky, K. Knappenberger, Jr., Hershberger, and A.W. Castleman, Jr. World Scientific, in press.
60. B.D. Leskiw, and A.W. Castleman, Jr. Clusters as Precursors of Nano-objects: Comptes Rendus de l'Académie des Sciences., submitted 2001.
61. S.E. Kooi and A.W. Castleman, Jr. J. Chem. Phys. **108**, 8864, (1998).
62. H. Sakurai, S.E. Kooi, and A.W. Castleman, Jr. J. Cluster Science, **10**, 493, (1999).
63. H. Sakurai and A.W. Castleman, Jr. J. Phys. Chem. **101**, 7695, (1997).
64. H. Sakurai and A.W. Castleman, Jr. J. Phys. Chem. A, **102**, 10486, (1998).
65. Sundstöm, V., ed. Femtochemistry and Femtobiology 1996, Imperial College Press, London.
66. Kelley, S.O.; Barton, J. K: Science **283**, 375, (1999).
67. Castleman, A.W., Jr.; Zhong, Q.; Hurley, S.M. Proc. Natl. Acad. Sci. **96**, 4219, (1999).
68. Kosower, E.M.; Huppert, D. Ann. Rev. Phys. Chem. **37**, 127, (1986).
69. Benesi, H.A.; Hildebrand, J.H. J. Am. Chem. Soc. **71**, 2703, (1949).
70. Mulliken, R.S. J. Am. Chem. Soc. **72**, 600, (1950).
71. Cheng, P.Y.; Zhong, D.; Zewail, A.H. J. Chem. Phys. **105**, 6216, (1996).
72. D. Zhong and A.H. Zewail, Proc. Natl. Acad. Sci. **96**, 2602, (1999).
73. Cheng, P.Y.; Zhong, D.; Zewail, A.H. Chem. Phys. Lett. **242**, 369, (1995).
74. DeBoer, G.; Burnett, J.W.; Fujimoto, A.; Young, M.A. J. Phys. Chem. **100**, 14882, (1996).
75. Su, J.T.; Zewail, A.H. J. Phys. Chem. A **102**, 4082, (1998).
76. Randall, K.L.; Donaldson, D.J. Chem. Phys. **211**, 377, (1996).
77. Fredin, L.; Nelander, B.J. Am. Chem. Soc. **96**, 1672, (1974).
78. Langhoff, C.A.; Gnädig, K.; Eisenthal, K.B. Chem. Phys. Lett. **46**, 117, (1980).
79. Hilinski, E.F.; Rentzepis, P.M. J. Am. Chem. Soc. **107**, 5907, (1985).
80. Lenderink, E.; Duppen, K.; Wiersma, D.A. Chem. Phys. Lett. **211**, 503, (1993).
81. Pullen, S.; Walker, L.A. II; Sension, R.J. J. Chem. Phys. **103**, 7877, (1995).
82. Lenderink, E.; Duppen, K.; Everdij, F.P. X.; Mavri, J.; Torre, R.; Wiersma, D.A. J. Phys. Chem. **100**, 7822, (1996).
83. Lee, J.; Griffin, R.D.; Robinson, G.W. J. Chem. Phys. **82**, 4920, (1985).
84. Pines, E.; Huppert, D.; Agmon, N.J. Chem. Phys. **88**, 5620, (1988).; Agmon, N.; Huppert, D.; Pines, E.J. Chem. Phys. **88**, 5631, (1988).
85. Smith, T.P.; Zakikia, K.Z.; Thakur, K.; Barbara, P.F. J. Am. Chem. Soc. **113**, 4035, (1991).

86. Brucker, G.A.; Kelley, D.F. J. Chem. Phys. 1989, 90, 5234; Brucker, G. A.; Kelley, D.F. Chem. Phys. Lett. **136**, 213, (1989).
87. Syage, J.A.; Steadman, J.J. Chem. Phys. **95**, 2497, (1991).
88. Kim, S.K.; Wang, J. -K.; Zewail, A.H. Chem. Phys. Lett. **228**, 369, (1994).
89. Syage, J.A. J. Phys. Chem. **99**, 5772, (1995).
90. Kim, S.K.; Breen, J.J.; Willberg, D.M.; Peng, L.W.; Heikal, A.; Sayage, J.A.; Zewail, A.H. J. Phys. Chem. **99**, 7421, (1995).
91. Knochenmuss, R. Chem. Phys. Lett. 1998, 293, 191; Knochenmuss, R. Chem. Phys. Lett. **311**, 439, (1999); Lührs, D.C.; Knochenmuss, R.; Fisher, I. Phys. Chem. Chem. Phys. **2**, 4335, (2000).
92. Knochenmuss, R.; Holtom, G.R.; Ray, D. Chem. Phys. Lett. 1993, 215, 188; Knochenmuss, R.; Smith, D.E. J. Chem. Phys. **101**, 7327, (1994).
93. Steadman, J.; Syage, J.A. J. Chem. Phys. **92**, 4630, (1990).
94. Syage, J.A. in Femtosecond Chemistry (Manz, J.; Wöste, L., Eds.) Springer-Verlag, Germany, pp. 475, (1994).
95. Pino, G; Grégoire, G.; Dedonder-Lardeux, C.; Jouvet, C.; Martrenchard, S.; Solgadi, D. Phys. Chem. Chem. Phys. **2**, 893, (2000); Grégoire, G.; Dedonder-Lardeux, C.; Jouvet, C.; Martrenchard, S.; Peremans, A.; Solgadi, D. J. Phys. Chem. A **104**, 9087, (2000).
96. Snyder, E.M.; Castleman, A.W., Jr. J. Chem. Phys. **107**, 744, (1997).
97. Wei S., Purnell J., Buzza S.A., Castleman A.W., Jr. J. Chem. Phys. **99**, 755, (1993).
98. Freudenberg, Th.; Radloff, W.; Ritze, H. -H.; Stert, V.; Weyers, K.; Noack, F.; Hertel, I.V. Z. Phys. D. **36**, 349, (1996).
99. Farmanara, P.; Radloff, W.; Stert, V.; Ritze, H.-H.; Hertel, I.V. J. Chem. Phys. **111**, 633, (1999).
100. Stert, V.; Radloff, W.; Freudenberg, Th.; Noack, F.; Hertel, I.V.; Jouvet, C.; Dedonder-Lardeux, C.; Solgadi, D. Europhys. Lett. **40**, 515, (1997).
101. C.A. Taylor, M.A. El Bayoumi, M. Kasha, Proc. Natl. Acad. Sci. **63**, 253, (1969).
102. K. Fuke, H. Yoshiuchi, K. Kaya, J. Phys. Chem. **88**, 5840, (1984).
103. K. Fuke, K. Kaya, J. Phys. Chem. **93**, 614, (1989).
104. A. Nakajima, F. Ono, Y. Kihara, A. Ogawa, K. Matsurbara, K. Ishikawa, M. Baba, K. Kaya, Laser Chem. **15**, 167, (1995).
105. A. Nakajima, M. Hirano, R. Hasumi, K. Kaya, H. Watanabe, C.C. Carter, J.M. Williamson, T. Miller, J. Phys. Chem. **101**, 392, (1997).
106. A. Douhal, S.K. Kim, A.H. Zewail, Nature **378**, 260, (1995).
107. Share, P., Pereira, M., Sarisky, M., Repinec, S., Hochstrasser, R.M. J. Lumin. **48/49**, 204, (1991).
108. S. Takeuchi, T. Tahara, Chem. Phys. Lett. **277**, 340, (1997); Takeuchi, S., Tahara, T. J. Phys. Chem. A **102**, 7740, (1998).
109. T. Fiebig, M. Chachisvilis, M. Manger, A.H. Zewail, A. Douhal, I. Garcia-Ochoa, A. de la Hoz Ayuso, J. Phys. Chem A **103**, 7419, (1999).
110. P. Hobza, personal communication. ,
111. Douhal, A.; Guallar, V.; Moreno, M.; Lluch, J.M. Chem. Phys. Lett. **256**, 370, (1996).
112. Bacic, Z.; Miller, R.E. J. Phys. Chem. A **100**, 12945, (1996).
113. Gerber, R.B.; Maccoy, A.B.; Garcia-Vela, A. Ann. Rev. Phys. Chem. **45**, 275, (1994).

114. Wöste, L.Z. Phys. Chem. **196**, 1, (1996)
115. Valentini, J.J.; Cross, J.B. J. Chem. Phys. **77**, 572, (1982).
116. Potter, E.D.; Liu, Q.; Zewail, A.H. Chem. Phys. Lett. **200**, 605, (1992).
117. Liu, Q.; Wang, J.-K.; Zewail, A.H. Nature **364**, 427, (1993).
118. Wang, J.-K.; Liu, Q.; Zewail, A.H. J. Phys. Chem. **99**, 11309, (1995).
119. Lienau, C.; Zewail, A.H. J. Phys. Chem. **100**, 18629, (1996).
120. Liu, Q.; Wang, J.-K.; Zewail, A.H. J. Phys. Chem. **99**, 11321, (1995).
121. Alexander, M.L.; Levinger, N.E.; Johnson, M.A.; Ray, D.; Lineberger, W.C. J. Chem. Phys. **88**, 6200, (1988).
122. Papanikolas, J.M.; Gord, J.R.; Levinger, N.E.; Ray, D.; Vorsa, V.; Lineberger, W.C. J. Phys. Chem. **95**, 8028, (1991).
123. Papanikolas, J.M.; Vorsa, V.; Nadal, M.E.; Campagnola, P.J.; Bushenau, H.K.; Lineberger, W.C. J. Chem. Phys. **99**, 8733, (1993).
124. Vorsa, V.; Campagnola, P.J.; Nandi, S.; Larsson, M.; Lineberger, W.C. J. Chem. Phys. **105**, 2298, (1996).
125. Vorsa, V.; Nandi, S.; Campagnola, P.J.; Larsson, M.; Lineberger, W.C. J. Chem. Phys. **106**, 1402, (1997).
126. Sanov, A.; Sanford, T.; Nandi, S.; Lineberger, W.C. J. Chem. Phys. **111**, 664, (1999).
127. Greenblatt, B.J.; Zanni, M.T.; Neumark, D.M. J. Chem. Phys. **111**, 10566, (1999).
128. Greenblatt, B.J.; Zanni, M.T.; Neumark, D.M. J. Chem. Phys. **112**, 601, (2000).
129. Batista, V.S.; Coker, D.F. J. Chem. Phys. **106**, 7102, (1997).
130. Faeder, J.; Delaney, N.; Maslen, P. E; Parson, R. Chem. Phys. Lett. **270**, 196, (1997).
131. Delaney, N.; Faeder, J.; Parson, R. J. Chem. Phys. **111**, 651, (1999).
132. Zadoyan, R.; Li, Z.; Ashjian, P.; Martens, C.C.; Apkarian, V.A. Chem. Phys. Lett. **218**, 504, (1994).
133. Apkarian, V.A. J. Chem. Phys. **101**, 6648, (1994).
134. Martens, C.C. J. Phys. Chem. **99**, 7453, (1995).
135. Poth, L.; Zhong, Q.; Ford, J.V.; Castleman, A.W., Jr. J. Chem. Phys. **109**, 4791, (1998).
136. Zhong; D.; Cheng, P.Y.; Zewail, A.H. J. Chem. Phys. **105**, 7864, (1996).
137. Donaldson, D.J.; Child, M.S.; Vaida, V. J. Chem. Phys. **88**, 7410, (1988).
138. Donaldson, D.J.; Vaida, V.; Naaman, R. J. Chem. Phys. **87**, 2522, (1987).
139. Vaida, V.; Donaldson, D.J.; Sapers, S.P.; Naaman, R.; Child, M.S. J. Phys. Chem. **93**, 513, (1989).
140. Kreisle, D.; Echt, O.; Knapp, M.; Recknagel, E.; Leiter, K.; Märk, T.D.; Sáenz, J.J.; Soler, J.M. Phys. Rev. Lett. **56**, 1551, (1986).
141. Bréchignac, C. Cahuzae, Ph.; Carlier, F.; de Frutos, M. Phys. Rev. Lett. **72**, 1636, (1994).
142. Shukla, A.K.; Moore, C.; Stace, A.J. Chem. Phys. Lett. **109**, 324, (1984); Stace, A.J. Phys. Rev. Lett. **61**, 306, (1988).
143. Kreisle, D.; Leiter, K.; Echt, O.; Märk, T.D. Z. Phys. D **3**, 319, (1986).
144. Scheier, P.; Märk, T.D. J. Chem. Phys. **86**, 3056, (1987); Lezius, M.; Märk, T.D. Chem. Phys. Lett. **155**, 496, (1989).
145. Scheier, P.; Stamatovic, A.; Märk, T.D. J. Chem. Phys. **88**, 4289, (1988); Scheier, P.; Dunser, B.; Märk, T.D. Phys. Rev. Lett. **74**, 3368, (1995).

146. Codling, K.; Frasinski, L.J.; Hatherly, P.A. J. Phys. B **21**, L433, (1988). (b) Codling, K.; Frasinski, L.J. J. Phys. B **26**, 783, (1993).
147. Normand, D.; Cornaggia, C.; Lavancier, J.; Morellec, J.; Liu, H.X. Phys. Rev. A **44**, 475, (1991).
148. Cornaggia, C.; Schmidt, M.; Normand, D. ., J. Phys. B: At. Mol. Opt. Phys. **27**, L123, (1994).
149. Wei, S.; Purnell, J.; Buzza, S.A.; Snyder, E.M.; Castleman, A.W., Jr. in Femtosecond Chemistry (Manz, J.; Wöste, L., Eds.) Springer-Verlag, Germany, pp. 449, 1994.
150. Snyder, E.M.; Wei, S.; Purnell, J.; Buzza, S.A.; Castleman, A.W., Jr. Chem. Phys. Lett. **248**, 1, (1996).
151. Purnell, J.; Snyder, E.M.; Wei, S.; Castleman, A.W., Jr. Chem. Phys. Lett. **229**, 333, (1994).
152. Snyder, E.M.; Buzza, S.A.; Castleman, Jr., A.W. Phys. Rev. Lett. **77**, 3347, (1996).
153. Ditmire T., Zweiback J., Yanovsky V.P., Cowan T.E., Hays G., Wharton K. B. Nature **398**, 489, (1999).
154. Ditmire, T.; Tisch, J.W. G.; Springate, E.; Mason, M.B.; Hay, N.; Smith, R.A.; Marangos, J.P.; Hutchinson, M.H. R. Nature **386**, 54, (1997).
155. Ford, J.V.; Zhong, Q.; Poth, L.; Castleman, Jr., A.W. J. Chem. Phys. **110**, 6257, (1999).
156. Ford, J.V.; Poth, L.; Zhong, Q.; Castleman, Jr., A.W. Int. J. Mass Spectrom. Ion Processes, **192**, 327, (1999).
157. McPherson, A.; Luk, T.S.; Thompson, B.D.; Boyer, K.; Rhodes, C.K. Appl. Phys. B **57**, 337, (1993).
158. Boyer, K., Thompson, B.D., McPherson, A., and Rhodes, C.K., J. Phys. B: At. Mol. Opt. Phys. **27**, 4373, (1994).
159. McPherson, A.; Luk, T.S.; Thompson, B.D.; Borisov, A.B.; Shiryaev, O. B.; Chen, X.; Boyer, K.; Rhodes, C.K. Phys. Rev. Lett. **72**, 1810, (1994).
160. McPherson, A.; Thompson, B.D.; Borisov, A.B.; Boyer, K.; Rhodes, C.K. Nature **370**, 631, (1994).
161. L. Koeller, M. Schumacher, J. Koehn, S. Teuber, J. Tiggesbaeumker, K.H. Meiwes-Broer, Phys. Rev. Lett. **82**, 3783, (1999).
162. Chelkowski, S.; Bandrauk, A.D. J. Phys. B: At. Mol. Opt. Phys. **28**, L723, (1995).
163. Zuo, T., Bandrauk, A.D., Phys. Rev. A **52**, R2511, (1995).
164. Chelkowski, S.; Conjusteau, A; Zuo, T.; Bandrauk, A.D. Phys. Rev. A **54**, 3235, (1996).
165. Yu, H.; Bandrauk, A.D. Phys. Rev. A **56**, 685, (1997).
166. Yu, H.; Bandrauk, A.D. J. Phys. B: At. Mol. Opt. Phys. **31**, 1533, (1998).
167. Seideman, T.; Ivanov, M.Y.; Corkum, P.B. Phys. Rev. Lett **75**, 2819, (1995).
168. Constant, W.; Stapelfeldt, H.; Corkum, P.B. Phys. Rev. Lett. **76**, 4140, (1996).
169. Rose-Petruck, C.; Schafer, K.J.; Barty, C.P. J. in Application of Laser Plasma Radiation II, Richardson, M.C.; Kyrala, G.A. (ed.) (SPIE, Bellingham, 1995), 2523, 272.
170. Rose-Petruck, C.; Schafer, K.J.; Wilson, K.R.; Barty, C.P. J. Phys. Rev. A **55**, 1182, (1997).
171. Last, I.; Schek, I.; Jortner, J.J. Chem. Phys. **107**, 6685, (1997).
172. Last, I.; Jortner, J. Phys. Rev. A **58**, 3826, (1998).

173. Last, I.; Jortner, J. Phys. Rev. A **62**, 13201, (2000).
174. Card, D.A.; Folmer, D.E.; Sato, S.; Buzza, S. A; Castleman, Jr., A.W. J. Phys. Chem. **101**, 3417, (1997).
175. Poth, L.; Castleman, Jr., A.W. J. Phys. Chem A **102**, 4075, (1998).
176. McPherson, A.; Boyer, K.; Rhodes, C.K. J. Phys. B: At. Mol. Opt. Phys. **27**, L637, (1994).
177. Chen, Z.Y., Walder, G.J., Castleman, Jr., A.W., J. Phys. Chem. **96**, 9581, (1992).
178. Jena, P., Khanna, S.N., Rao, B.K., in: Kumar, V., Martin, T.P., Tosatti, E. (Eds.) Clusters and Fullerenes, World Scientific 1992, p. 73.
179. Reddy, B.V., Khanna, S.N., Chem. Phys. Lett. **209**, 104, (1993).
180. Li, Z-Q., Gu, B.-L., Han, R.-S., Zheng, Q.-Q., Z. Phys. D. **27**, 275, (1993).
181. Lou, L., Guo, T., Nordlander, P., Smalley, R.E., J. Chem. Phys. **99**, 5301, (1993).
182. Hay, P.J., J. Phys. Chem. **97**, 3081, (1993).
183. Khan, A., J. Phys. Chem. **97**, 10937, (1993).
184. Lin, Z., Hall, M.B., J. Am. Chem. Soc. **115**, 11165, (1993).
185. Rohmer, M.-M., Bénard, M., Poblet, J.-M., Chem. Rev. **100**, 495, (2000).
186. Leisner, T., Athanassenas, K., Echt, O., Kreisle, D., Rechnagel, E., J. Chem. Phys. **99**, 9670, (1993).
187. Leisner, T., Athanassenas, K., Echt, O., Kandler, O., Kreisle, D., Rechnagel, E., Z. Phys. D **20**, 127, (1991).
188. Collings, B.A., Amrein, A.H., Rayner, D.M., Hackett, P.A., J. Chem. Phys. **99**, 4174, (1993).
189. Nieman, B.C., Parks, E.K., Richtsmeier, S.C., Liu, K., Pobo, L.G., Riley, S.L., High Temp. Sci. **22**, 115, (1986).
190. Athanassenas, K., Leisner, T., Frenzel, U., Kreisle, D., Ber. Bunsenges. Phys. Chem. **96**, 1192, (1992).
191. Amrein, A., Simpson, R., Hackett, P., J. Chem. Phys. **95**, 1781, (1991).
192. Campbell, E.E.B., Ulmer, G., Hertel, I.V., Phys. Rev. Lett. **67**, 1986, (1991).
193. Ding, D., Huang, J., Compton, R.N., Klots, C.E., Haufler, R.E., Phys. Rev. Lett. **73**, 1084, (1994).
194. Lin, H., Han, K.-L., Bao, Y., Gallogly, E.B., Jackson, W.M., J. Phys. Chem. **98**, 12495, (1994).
195. Klots, C.E., Chem.Phys. Lett. **186**, 73, (1991).
196. Leskiw, B.D., Knappenberger, Jr., K.L., Castleman, Jr., A.W., in preparation.
197. Chen, J.G. De Vries, B.D. Frühberger, B., Kim, C.M., Liu. Z.-M., J. Vac. Sci. Technol. A **13**, 1600, (1995).
198. Heijnsbergen, D. Helden, G., Duncan, M.A., Roij, A.J.A., Meijer, G., Phys. Rev. Lett. **83**, 4983, (1999).

10 Future Directions

A.W. Castleman, Jr. and S.N. Khanna

The present book has shown how clusters and nanostructured materials could serve as the laboratories for observing quantum phenomenon and how the quantum phenomenon can modify the behavior at reduced sizes and scale. There are however, outstanding problems that still remain unsolved. In the area of thermal properties, the variation of the melting point, changes in entropy at the melting and the negative heat capacity are still not understood. In fact, the question of phase transition in reduced sizes and how to extend the classical theory of thermodynamics are still challenging problems.

There is extensive interest in probing the electronic, optical and reactive behavior of clusters. One of the recent areas is probing the dynamics of quantum electronic states in these systems. In this regard the recent application of femtosecond pump-probe techniques to explore the nature of the electronic states including the excitation and relaxation dynamics is providing information into the evolution of the band structure and the mechanism for chemisorptions. Indeed, the study of cluster reactions is a rapidly growing area of research which is providing new insights into the fundamentals of solvation effects on reaction dynamics, as well as into developments that may eventually yield methods to tailor the design of nanocatalysts with a high degree selectivity. We are already seeing the emergence of results that are giving a deeper understanding of reactions of biological significance. There is a great need for theoretical techniques that can complement these experiments and thus provide insights into the temporal evolution of intermediate and product states.

Magnetic phenomena in reduced sizes are probably among the most unique. Recent studies have shown that molecular nanomagnets exhibit quantum steps in the hysterisis loops. These steps are reminiscent of the occupation of the states arising due to angular momentum quantization and hence provide us with unique opportunity to observe quantum effects at the macroscopic scale. The experimental data on relaxation can be analyzed in terms of the model hamiltonians and a detailed understanding of these is still lacking. Note that the magnetic anisotropy is probably the single most important quantity in the magnetic phenomenon and a microscopic understanding of the various factors that constitute the magnetic anisotropy is still lacking. Part of the difficulty is that an exact treatment of the spin orbit coupling requires inclusion of relativistic effects. This is both numerically difficult and requires development of appropriate functions. Another important class of problems is the magnetic behavior of clusters supported on molecular templates and the effect of adsorbates on the magnetic moments and the dynamical behavior.

Reduction in device sizes has pushed the frontiers of condensed matter physics to the molecular level. It is then possible to envision integrated circuits that will be several thousand times denser and higher in speed. It is expected that the present day silicon technology will end in about 10 years with a limited size of 0.1 mm. The newer devices would be composed of molecular units and a case in point is a recent suggestion of a molecular unit for giant magnetoresistance. For example, conducting molecules can be sandwiched between molecular magnets. One can then create giant magnetoresistance by aligning or misaligning the local magnetic moments of the molecular magnets. At the same time, it is envisioned that one can create molecular units composed of transition metal atoms encapsulated in semiconducting cages. If the magnetic moment of the transition metal atom could be stabilized, this would provide a unique opportunity to create devices where one could simultaneously use the charge and spin of the electron. This new area on "spintronix" is currently extremely active and will be very important for putting the nanodevices into applications.

Using the photodetachment spectroscopy, it now seems possible to probe the molecular rearrangement dynamics in complex molecules. Further, the electronic states can be probed by negative ion photodetachment spectroscopy. When combined with accurate theoretical calculations, this could provide information on the spin magnetic moment and the metal insulator transitions. Information on the spin magnetic moment is quite important as it is difficult to determine magnetic moments of small clusters using the conventional Spin-Gerlach experiments. Another area relates to the formation of special class of negative ions by binding electrons to molecules having large dipole moments, termed dipole bound states. The molecules are stabilized by the transfer of energy between the atoms in high Rydberg states and polar molecules. Experimental techniques are also allowing fabrication of binary clusters composed of aromatic systems. Many of these will allow us to understand dynamics in biological systems. For example, the solvation of K^+ ions by benzene can be used to derive information on the motion on these ions in human beings and may have applications in understanding hyper-tension. The application of nanomaterials to biological systems is perhaps the most exciting and promising future area of research.

Upon considering the many developments surveyed in this book, we believe the reader will find the subject to be a challenging and rich one. It is clear that the behavior of matter of nanoscale dimensions is generally not directly predictable from the known properties of the related bulk materials. The exciting prospects for further developments in the field are evident, as is realization that we are just beginning to see the opportunity offered for obtaining a deeper understanding of the unique properties of matter of restricted geometry and small size due to the constraints imposed by quantum mechanics. And in terms of applications, it is evident that through clusters, there is the alternative promising approach of assembling new materials of nanoscale dimensions from the "bottom up" rather than the "top down". The prognosis for the field is excellent, and we can expect to see many new and rapidly unfolding advances in the coming years. The field is certain to have a long and prosperous life. We hope the reader of this book will be stimulated to join in contributing to its further growth.

Index

7-Azaindole dimer   229

acetone clusters   242
Ag submonolayer structures   103
ammonia clusters   228
anticovariance map of ammonia clusters   239
Ar_{13} 16
Ar_{55} 19
atomic scale template   110
Au clusters   184
Au_3^- 50
Auger parameter   177
avoided curve-crossing   227
axial anisotropy   63

Barnett experiment   100
Berry phase   65
blocking temperature   129
Bz•I_2 system   225

C_{60} clusters   29
caging dynamics   224, 231
caging dynamics in anionic clusters   234
catalysis by supported metal clusters   159
CEMM model   241
chemical map   166
chemical reactivity   45
cluster dynamics   223
cluster magnetism   87
cluster size and reactivity   170
clusters on surfaces   117
CO chemisorption   46
CO dissociation   186
CO dissociation activity   187
CO oxidation on Au/TiO_2 185

CO oxidation over a Pt/MgO   189
CO oxidation turnover frequencies   185
CO oxidization on a Pt cluster   191
Co/Au(111)   115
Co grown on Au(111)   115
Co on Au(111)   121
cobalt islands   104
coherent dissociation   233
confined nucleation   113
core levels shifts   176
Coulomb blockage   182
Coulomb explosion   224, 238

decay processes in C_{60} 212
delayed ionisation from C_{60} 215
delayed ionisation tail from C_{60} 210
delayed ionization   199
demagnetized σ_z maps   127
detailed balance   203
diffusion limited aggregation model (DLA)   104
dodecanuclear manganese clusters   77
double proton-transfer   229
Dzyaloshinsky-Moriya term   58

Einstein-de Haas experiment   100
electron-transfer reactions   224
emission spectra   176
exchange coupled clusters   120
excited-state proton-transfer   226

F-centers   189
Fe clusters   174
Fe clusters on W(110)   128
Fe on Cu/Pt(111)   113
Fe stripes on vicinal Cu(111)   122
Fe_8Br clusters   56, 74
$[Fe_8O_2(OH)_{12}(tacn)_6]Br_89H_2O$ 56

Franck-Condon transition 34
Frenkel-Kontorova model 109
fs excitation 219

Gd_N clusters 97
giant moments 93
gold catalyst 185
growth kinetics of clusters 102

heats of deprotonation 226
Heisenberg model 86
helium clusters 11
hexagonal tower 151
HFEPR 67
hole digging 72
hole digging method 71
Hund's rules 87

I_2 232
$I_2^-(CO_2)_n$ 236
interacting islands 120
interatomic distances of Pt clusters 172
Island ordering 114

Jellium model 3

$(KCl)_{32}$ 16
kinetic model 217
Kubo's approximation 179

Landau-Zener model 65
ligand-stabilized clusters 163
locked magnetic moment 97, 118

magic numbers 13
magnetic anisotropy 57, 100, 154
Met-Cars 245
– $Ti_8C_{12}^+$ 246
– Ti_8C_{12} 247
– $Ti_xNb_yC_{12}$ 247
– $Ti_xZr_yC_{12}$ 247
– V_8C_{12} 247
– Zr_8C_{12} 247
metal-to-nonmetal transition 178
$Mn_{12}O_{12}$ 140
$Mn_{12}O_{12}(RCOO)_{16}(H_2O)_4$ cluster 148
$(MnO)_3$ rings 149

$[Mn_{12}O_{12}(CH_3COOH)_{16}(H_2O)_4]$ 56
molecular dynamics 8

nano-lithography 164
nanocluster catalysts 192
Ni_nH_m clusters 143
non-exponential relaxation 69
nucleation on ordered dislocations 112

Orbach mechanism 61
orbital magnetic moment 119
organized growth 108

PES^- of Cu_1^- 34
phase diagrams 19
phase equilibrium 13
photoelectron spectra
– Ag_7^- 41
– Ag_8^- 41
– Au_2^- 50
– Au_3^- 50
– Hg_n^- clusters 43
– Mg_n 42
– Na_7^- 40
– $Ti_nH_m^-$ clusters 48
photoelectron spectra from Na_{93}^+ 217
photoelectron spectroscopy 31
– configuration interaction 33
– multiplet splitting 32
– relaxation 32
– shake up process 33
– single particle picture 33
Pt atoms on Pt(111) 104
Pt cluster on $TiO_2(110)$ 167
Pt nanoclusters 164
Pt/titania catalyst 169
Pt_8 clusters 189
pump-probe photoelectron spectrum 50
pump-probe spectroscopy 223

quantum tunneling 55, 146

rate constant 206
rate constant for ionization 203
reactivity of Au/TiO_2 181
reactivity of stepped surfaces 187
recombination dynamics 233

resonance multiphoton ionization (REMPI) 225
Rh_N clusters 97
Rh carbonyls 180
Rh island 188
$Rh/Al_2O_3/NiAl$ (110) 186
Rydberg states 219

scanning tunneling spectroscopy 182
self organization 112
Si_n^- clusters 43
solvation effect of water clusters 230
spectromicroscopy 166
spectroscopy of clusters 29
spin canting 97
spin-orbit coupling 155
spin-phonon interaction 61
SPLEED-polarization 129
stepped hysteresis 63
Stern-Gerlach experiments 83, 89
Stranski-Krastanov mode 106
STS curves 184
substrate polarization 177

superparamagnetic (SP) relaxation 141
superparamagnetic behavior 55, 61, 91

tautomerization reaction 243
thermal excitation 219
thermionic emission 199
$Ti_nH_m^-$ clusters 46
TiO_2 surface upon Au deposition 162
tip proximity effects 183
TPR spectra 189
transition state theory 200
tunnel splitting in Fe_8Br 68

ultrafast spectroscopy 248

vapor deposition 161
Volmer-Weber growth mode 107, 172, 182

zero delay 50

Printing (Computer to Plate): Saladruck Berlin
Binding: Stürtz AG, Würzburg

Springer Series in
CLUSTER PHYSICS

Editor-in-Chief:

Professor A. W. Castleman, Jr.
Department of Chemistry, The Pennsylvania State University
152 Davey Laboratory, University Park, PA 16802, USA

Editorial Board:

Professor R. Stephen Berry
Department of Chemistry, The University of Chicago
5735 South Ellis Avenue, Chicago, IL 60637, USA

Professor Dr. Hellmut Haberland
Albert-Ludwigs-Universität Freiburg, Fakultät für Physik
Hermann-Herder-Strasse 3, 79104 Freiburg, Germany

Professor Dr. Joshua Jortner
School of Chemistry, Tel Aviv University
Raymond and Beverly Sackler, Faculty of Sciences
Ramat Aviv, Tel Aviv 69978, Israel

Dr. Tamotsu Kondow
Toyota Technological Institute, Cluster Research Laboratory
East Tokyo Laboratory, Genesis Research Institute Inc.
Futamata 717-86, Ichikawa, Chiba 272-0001, Japan

You are one click away from a world of physics information!

Come and visit Springer's
Physics Online Library

Books
- Search the Springer website catalogue
- Subscribe to our free alerting service for new books
- Look through the book series profiles

You want to order? Email to: orders@springer.de

Journals
- Get abstracts, ToC´s free of charge to everyone
- Use our powerful search engine SpringerLink Search
- Subscribe to our free alerting service SpringerLink *Alert*
- Read full-text articles (available only to subscribers of the paper version of a journal)

You want to subscribe? Email to: subscriptions@springer.de

Electronic Media
- Get more information on our software and CD-ROMs

You have a question on an electronic product? Email to: helpdesk-em@springer.de

•••••••••••• Bookmark now:

www.springer.de/phys/

Springer · Customer Service
Haberstr. 7 · D-69126 Heidelberg, Germany
Tel: +49 (0) 6221 345 - 0 · Fax: +49 (0) 6221 345-4229
d&p · 006437_sf1c_1c